science for common entrance

ISEB
Independent Schools
Examinations Board

Peter D Riley

editors
Richard Balding
David Penter

HODDER
EDUCATION
PART OF HACHETTE LIVRE UK

Every effort has been made to trace all copyright holders, but if any have been inadvertently overlooked the Publishers will be pleased to make the necessary arrangements at the first opportunity.

Hachette Livre UK's policy is to use papers that are natural, renewable and recyclable products and made from wood grown in sustainable forests. The logging and manufacturing processes are expected to conform to the environmental regulations of the country of origin.

Orders: please contact Bookpoint Ltd, 130 Milton Park, Abingdon, Oxon OX14 4SB. Tel (44) 01235 827720. Fax (44) 01235 400454. Lines are open 9.00–5.00, Monday to Saturday, with a 24-hour message answering service. You can also visit our website at www.hoddereducation.co.uk.

© Peter Riley, Richard Balding, David Penter 2005
First published in 2005 by
Hodder Education,
part of Hachette Livre UK
338 Euston Road
London NW1 3BH

Impression number   10 9 8 7 6 5 4 3 2
Year   2010 2009 2008

Cover design John Townson/Creation
Typset in 11.5/14pt Minion by Pantek Arts Ltd, Maidstone, Kent
Printed in Italy

A catalogue record for this title is available from the British Library

ISBN-10: 0340 889 411
ISBN-13: 978 0340 889 411

Contents

To the pupil

If you watch a baby playing with a rattle, you will see how the senses of sight, touch and even taste are used in an investigation. No doubt the senses of hearing and smell are being used too. From our earliest days we are natural investigators. We investigate the world around us to help us understand it.

When you began school, you started doing scientific investigations. These involved making observations, thinking of explanations for what you saw, trying experiments to test your explanations, and comparing your results with your ideas. This way of investigating is called the scientific method and has been used by scientists for hundreds of years.

A vast amount of information has been built up by scientific investigations. To make it easier to study, this information has been divided up into three areas of study called biology, chemistry and physics, as you will find in this book.

Biology is the scientific study of living things. It includes investigations on tiny structures, such as cells, and on huge structures such as a rainforest, an ocean or even the whole Earth! Some biologists are even looking for signs of life in other parts of the Solar System or on planets around other stars.

Chemistry is the scientific study of matter and materials – the substances such as air, water and rock, which make up our world. Matter is made from tiny structures called atoms that join together in many different ways to make millions of different kinds of chemicals. The first atoms formed at the beginning of the Universe. At that time a huge explosion called the Big Bang took place. The first atoms formed two gases – hydrogen and helium. Over thousands of millions of years since then, many other atoms have been made in the stars. When the stars faded out or exploded in a supernova, these atoms spread through space and formed dust. In time, one cloud of dust formed the Solar System and everything that is in it. This means that all the atoms in the book you are holding now once formed in the Universe millions of years ago. It also means that the atoms that make up your hands formed then.

Physics is the scientific study of the interactions between matter and energy. These interactions can produce the colours of the rainbow in a shower or the roar of the wind in a hurricane. At a greater distance, the interactions of matter and energy in the Sun produce light and heat, while inside our eyes light energy is converted into electrical energy, which passes to our brain and allows us to see. Every event in the Universe, from your next breath to a star exploding, is an interaction of matter and energy, so physics is really a part of all other scientific subjects, rather than a separate one.

Science for Common Entrance covers the requirements for your examination in a way that I hope will help you understand how observations, ideas and investigations have led to the scientific facts we use today. The questions are set to help you extract information from what you read and see, and to help you think more deeply about each chapter in this book. Some questions are set so that you can discuss your ideas with others and develop a point of view on different scientific issues. This should help you in the future when new scientific issues, which are as yet unknown, affect your life.

The scientific activities of thinking up ideas to test and carrying out investigations are enjoyed so much by many people that they take up a career as scientists. Perhaps *Science for Common Entrance* may help you take up a career in science too.

To the teacher

Science for Common Entrance has been developed from the second editions of the three books *Biology Now! 11–14*, *Chemistry Now! 11–14* and *Physics Now! 11–14*, to cover the requirements of the Common Entrance Examination for Science. A variety of questions are set in each chapter to test a range of intellectual competencies as set out in Bloom's Taxonomy and briefly explained in the accompanying *Teacher's Resource Book* (see below). There are also 'For discussion' questions which aim to encourage the pupils to talk about a particular area of study or wider issues to which science makes a contribution. These questions provide an opportunity to build up pupils' scientific literacy skills, and some questions can also be used in work on citizenship.

The *Teacher's Resource Book* provides support for each chapter in *Science for Common Entrance*. The answers to all the questions in the chapters are given – those that occur in the body of the text and those that occur at the end of each chapter, which can be used for extra assessment. A range of practical activities, some including worksheets, are provided in the Resource Book for integration with the content of each chapter in the Pupil's Book, to provide opportunities for pupils to develop their skills of investigation.

There are also suggestions for integrating ICT into each chapter. There are past examination questions and mark schemes to help the pupils prepare for the examination, and there is a table in which the requirements of the Common Entrance curriculum are set out with page references to the Pupil's Book, to help you track curriculum coverage.

Acknowledgements

I would like to thank Richard Balding, David Penter and Jane Roth for their editorial work in the development of this book.

Thanks to FRANK for use of the drugs advice poster in this book (Figure 10.10 **p. 100**). FRANK is available 24 hours every day for confidential advice and information about drugs, 0800 77 66 000 or visit talktofrank.com.

The following are sources from which diagrams and data have been adapted:
Figure 17.5 **p. 155** from *Letts Key Stage 3 Study Guide: Science* by B. McDuell & G. Booth, by permission of Letts Educational.
Table 21.1 **p. 192** from Figure 6.2 of *Chemistry Today* by Euan Henderson, Macmillan, with permission.
Tables 24.2 & 24.3 **p. 244** from R. S. Holt, *School Science Review* September 1980, 62 (218), with permission.
Table 28.1 **p. 287** from M. S. Byrne, *School Science Review* June 1981, 62 (221), p. 749, with permission.

The following have supplied photographs or given permission for photographs to be reproduced:
Cover John Townson/Creation; **p.2** *tl, r* Heather Angel/Natural Visions, *bl* Gerard Lacz/NHPA; **p.6** J. C. Revy/Science Photo Library; **p.10** Carolina Biological Supply Co./Oxford Scientific Films; **p.15** © Tom Galliher/CORBIS; **p.16** Andrew Lambert; **p.20** CNRI/Science Photo Library; **p.21** Motta & Familiari/Anatomy Dept/University 'La Sapienza' Rome/Science Photo Library; **p.24** *l* Sally Greenhill © Sally & Richard Greenhill, *r* Andrew Lambert; **p.27** David Woodfall/NHPA; **p.29** David Woodfall/NHPA; **p.30** *t* John Hawkins/Frank Lane Picture Agency, *b* Stephen Dalton/NHPA; **p.31** *l* Neil McIntyre, *r* Allan G. Potts/Bruce Coleman Collection; **p.35** *l* Heather Angel/Natural Visions, *c* Andrew Henley/Natural Visions, *r* Jany Sauvanet/NHPA; **p.36** Sally Greenhill © Sally & Richard Greenhill; **p.43** Andrew Lambert; **p.45** Heather Angel/Natural Visions; **p.63** *t* Cordelia Molloy/Science Photo Library, *b* John Townson/Creation; **p.66** WaterAid/Caroline Penn; **p.70** Harwood/Ecoscene; **p.71** *t* Sally Morgan/Ecoscene, *b* © Jacomina Wakeford/ICCE; **p.77** Richard Davies/ Oxford Scientific Films; **p.78** Wildlife Matters; **p.79** Alexandra Jones/ Ecoscene; **p.80** *t* Garden & Wildlife Matters, *l* John Mason/Ardea   London Ltd; **p.85** A. B. Dowsett/Science Photo Library; **p.88** *t* John Townson/Creation, *bl* Holt Studios/Bob Gibbons, *br* Holt Studios/Nigel Cattlin; **p.93** Deep Light Productions/Science Photo Library; **p.100** John Townson/Creation; **p.104** Heather Angel/Natural Visions; **p.107** Harry Smith Horticultural Photographic Collection; **p.109** *all* Holt Studios/Nigel Cattlin; **p.112** Stefan Meyers/Ardea London Ltd; **p.114** *all* John Townson/Creation; **p.117** Holt Studios/Nigel Cattlin; **p.120** *all* John Townson/Creation; **p.122** *l* Science Museum/Science & Society Picture Library, *r* Geoff Tompkinson/Science Photo Library; **p.123** *both* Andrew Lambert; **p.124** Andrew Lambert; **p.125** Andrew Lambert; **p.126** *all* Andrew Lambert; **p.127** Andrew Lambert;

p.131 *c* Peter Menzel/Science Photo Library, *l* John Townson/Creation; **p.132** Dex Image/Alamy; **p.133** Tim Mosenfelder/CORBIS; **p.135** Heather Angel/Natural Visions; **p.136** *tl* John Townson/Creation, *c* Andrew Lambert; **p.138** *all* Andrew Lambert; **p.140** *l, r* John Townson/Creation, *c* Phil Chapman; **p.141** Andrew Lambert; **p.142** *all* John Townson/Creation; **p.143** *lt* Pete Atkinson, *lb* Ken Lucas/Ardea, *b* Andrew Lambert; **p.145** *t* John Townson/Creation, *c* Ecoscene/Gryniewicz, *b* Andrew Lambert; **p.148** *all* Andrew Lambert; **p.149** *l* Peter Knab/Sainsbury's '*The Magazine*', *r* John Townson/Creation; **p.153** *t* David Woodfall/NHPA, *b* Biophoto Associates/Science Photo Library; **p.154** *all* Andrew Lambert; **p.157** John Townson/Creation; **p.158** *both* Andrew Lambert; **p.161** *both* John Townson/Creation; **p.165** Andrew Lambert; **p.167** Science Photo Library; **p.168** *both* Geoscience Features Picture Library; **p.169** Geoff Tompkinson/Science Photo Library; **p.170** *t* Paul Brierley Photo Library, *bl & br* ILFORD Imaging; **p.172** *all* John Townson/Creation; **p.174** Juhan Kuus/Rex Features; **p.175** *t* Mary Evans Picture Library, *cl & cr* Andrew Lambert; **p.176** John Townson/Creation; **p.177** Andrew Lambert; **p.179** *t* Tek Image/Science Photo Library, *b* Flip Schulke; **p.180** *c* Mark Brewer/Rex Features, *l* David Parker/Science Photo Library; **p.183** *both* Geoscience Features Picture Library; **p.184** *both* John Townson/Creation; **p.185** *all* John Townson/Creation; **p.186** John Townson/Creation; **p.189** John Townson/Creation; **p.190** Andrew Lambert; **p.191** *all* Andrew Lambert; **p.192** Andrew Lambert; **p.193** *all* Geoscience Features Picture Library; **p.194** Gerald Cubitt/Bruce Coleman Collection; **p.195** *t* Pirelli Cables Ltd, *l* Ronald Sheridan/Ancient Art & Architecture Collection; **p.197** Jaguar Cars Ltd; **p.203** *t* Mary Evans Picture Library, *b* Giles Angel/Natural Visions; **p.206** *t* Heather Angel/Natural Visions, *l* Hulton Archive; **p.207** Robert Harding Picture Library; **p.208** Astrid & Hanns-Frieder Michler/Science Photo Library; **p.209** *all* John Townson/Creation; **p.210** *l* Guy Edwardes/NHPA, *c* Simon Fraser/Science Photo Library; **p.211** *t* Geoscience Features Picture Library, *b* Ecoscene/Alexandra Jones; **p.212** Ecoscene/Kieran Murray; **p.213** *t* Ecoscene/Sally Morgan, *b* Ecoscene/W. Lawler; **p.214** Ecoscene/Chinch Gryniewicz; **p.215** John Townson/Creation; **p.218** China Great Wall Industry Corp./Science Photo Library; **p.220** David Nunuk/Science Photo Library; **p.222** Keith Kent/Science Photo Library; **p.225** Andrew Lambert; **p.229** *all* John Townson/Creation; **p.231** Space Telescope Science Institute/NASA/Science Photo Library; **p.232** Will Blanche/Rex Features; **p.234** *lt* John Townson/Creation, *lc* Action Images, *b* Action Images/Bob Martin; **p.236** *t* NASA/Science Photo Library, *b* Hartmut Schwarzbach/Still Pictures; **p.237** *t & br* Mark Edwards/Still Pictures, *bl* Jorgen Schytte/Still Pictures; **p.238** *cl* Andrew Lambert, *cr* Richard Folwell/Science Photo Library, *b* NASA/Science Photo Library; **p.239** *c* Klaus Andrew/Still Pictures, *b* Novosti Press Agency/Science Photo Library; **p.240** *t* Ocean Power Delivery Ltd, *c* Martin Bond/Science Photo Library; **p.241** *tl* Simon Fraser/Science Photo Library, *c* Robert Francis/Robert Harding Picture Library, *bl & br* Greg Balfour/Evans/Greg Evans International; **p.243** Alstom; **p.246** John Townson/Creation; **p.251** Andrew Lambert; **p.252** *c* Maplin, *b* Andrew Lambert; **p.253** *t* Maplin, *b* Andrew Lambert; **p.254** *all* John Townson/Creation; **p.255** Alstom; **p.256** Martin

Bond/Science Photo Library; **p.258** *l*, *ct* & *rb* Action Images/Sporting Pictures (UK) Ltd, *rt* Robert Harding Picture Library, *cb* Adam Woolfit/Robert Harding Picture Library; **p.259** Action Images/Sporting Pictures (UK) Ltd; **p.260** *t* Isobel Cameron/Forest Life Picture Library © Crown, *b* Dr Jeremy Burgess/Science Photo Library; **p.261** Action Images/Sporting Pictures (UK) Ltd; **p.262** Alex Bartel/Science Photo Library; **p.265** Andrew Lambert; **p.267** National Meteorological Office © Crown; **p.268** NASA/Science Photo Library; **p.269** NASA/Science Photo Library; **p.272** Jerry Schad/Science Photo Library; **p.276** George East/ Science Photo Library; **p.279** Dr Fred Espenak/Science Photo Library; **p.280** Jerry Lodriguss/Science Photo Library; **p.283** *both* Andrew Lambert; **p.288** Andrew Lambert; **p.289** *both* Andrew Lambert; **p.292** *c* Alex Bartel/Science Photo Library, *b* Maplin; **p.293** Maplin; **p.297** Nigel Francis/Robert Harding Picture Library; **p.298** Gordon Garradd/Science Photo Library; **p.301** *t* Matthias Breiter/Oxford Scientific Films, *c* John Townson/Creation, *b* Derek Cattani; **p.307** *l* Alexis Rosenfeld/Science Photo Library, *r* Xavier Eichaker/Still Pictures; **p.310** H. Verbiesen/Still Pictures; **p.315** Jon Feingersh/CORBIS; **p.316** Keith Kent/Science Photo Library; **p.318** Takeshi Takahara/Science Photo Library; **p.319** *t* Action Images, *b* Detlev Van Ravenswaay/Science Photo Library; **p.320** *tl* Jeffrey Rotman/Still Pictures, *tr* Richard Herrmann/Oxford Scientific Films, *bl* Martyn F. Chillmaid/Robert Harding Picture Library, *br* Philip Dunn/Rex Features; **p.325** *t* Action Images, *b* Bildagentur Schuster/Robert Harding Picture Library; **p.326** John Townson/Creation; **p.329** Bildagentur Schuster/Herbst/ Robert Harding Picture Library.

t = top, *b* = bottom, *l* = left, *r* = right, *c* = centre

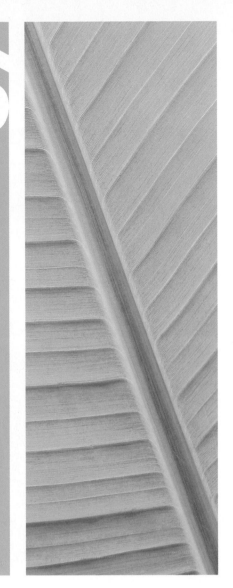

biology

The structure of living things

Figure 1.1   There is a great variety in the structure of living things.

---

1 Read the section *Organ systems of a human* on pages 2–4. Which organ system:

a) transports materials around the body,

b) absorbs food into the blood,

c) detects changes in the environment,

d) produces hormones,

e) co-ordinates activities,

f) takes in oxygen from the air,

g) supports the body,

h) produces offspring,

i) removes waste from the blood,

j) moves bones?

Organs and organ systems

Biology is the study of living things. All living things are made up of cells. Groups of similar cells form tissues of different types. The parts of a body that perform specific tasks to keep a living thing alive are called organs and most organs work together in groups called organ systems.

Organ systems of a human

There are ten organ systems in the human body. They are listed in this section but will be covered in more detail throughout the book. The tasks they carry out are sometimes called life processes.

1   The **sensory system** is made up of sense organs such as the eye and the ear. The function of this system is to provide information about the surroundings.

2   The **nervous system** comprises the brain, spinal cord and nerves. This system controls the actions of the body and co-ordinates many of its activities without you having to think about them. For example, you breathe in and out automatically.

2 Which organ system or systems are involved in:
  a) movement,
  b) nutrition,
  c) circulation?
3 What other sense organs are in the sensory system?
4 What movements take place in the body that you do not have to think about?
5 How does your pattern of breathing change when you exercise and then rest?
6 Write down the names of any bones that you know (without looking them up) and say where they are found in the body.

3 The **respiratory system** is located in the chest. It is formed by the windpipe, the lungs, the ribs and rib muscles (called the intercostal muscles) and the diaphragm. The system works to draw in air and then expel it. While in the lungs, oxygen passes from the air into the blood and carbon dioxide passes from the blood into the air.

4 The **digestive system** is a long tube through the body in which food is broken down and absorbed into the blood. It is made up of many organs including the liver, which also performs many other tasks in the body.

5 The **circulatory system** transports materials around the body in a liquid called the blood. Blood is moved along tubes called blood vessels by the pumping action of the heart.

6 The **excretory system** cleans waste from the blood by a filtration process in the kidneys. The liquid containing the waste is called urine. It is stored in the bladder before it is released.

7 The **skeletal system** is made up of 206 bones. They provide support for the body and have joints between them that help the body to move. Some bones form a protective structure, for example, the bones in the skull form a protective case around the brain.

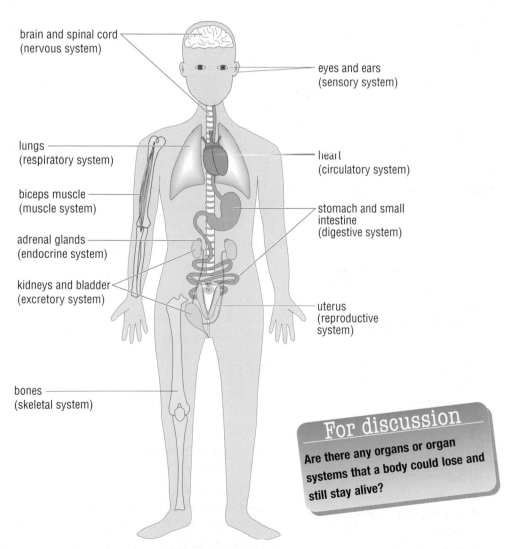

brain and spinal cord (nervous system)

eyes and ears (sensory system)

lungs (respiratory system)

heart (circulatory system)

biceps muscle (muscle system)

stomach and small intestine (digestive system)

adrenal glands (endocrine system)

kidneys and bladder (excretory system)

uterus (reproductive system)

bones (skeletal system)

For discussion

Are there any organs or organ systems that a body could lose and still stay alive?

Figure 1.2 Organs of the human body.

7 You are walking across a road and hear a sound behind you. You turn and see that a car has swerved to avoid a cat and is heading straight for you. What body systems work to get you out of the car's way? Why do you think these systems developed?

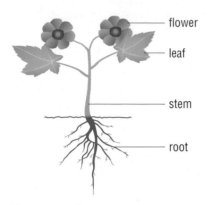

— flower

— leaf

— stem

— root

Figure 1.3 Organs of a flowering plant.

8 Draw a table featuring the organs of a flowering plant and the tasks they perform.
9 How is the leaf dependent on the root and the stem?
10 Which life processes or tasks do you think are found in both plants and humans? Explain your answer.

8 The **muscle system** provides the mechanism for movement. A muscle is capable of making itself shorter to exert a pulling force on a bone.
9 The **endocrine system** is made up of glands which release chemicals called hormones into the blood. Hormones control many processes in the body, including growth and sexual development.
10 The **reproductive system** of the male produces sperm cells and the reproductive system of the female produces eggs and provides a place for a baby to grow.

Organs of a flowering plant

There are four main organs in the body of a flowering plant. They are the root, stem, leaf and flower. Each organ may be used for more than one task or life process.

1 The **root** anchors the plant and takes up water and minerals from the soil. The roots of some plants, such as the carrot, store food.
2 The **stem** transports water and food and supports the leaves and the flowers. Some plants, such as trees, store food in their stems.
3 The **leaf** produces food. In some plants, such as the onion, food is stored in the bases of the leaves. The swollen leaf bases make a bulb.
4 The **flower** contains the reproductive organs of the plant.

All the organs work together to keep the plant alive so that it can grow and produce offspring.

The microscope

A microscope is used for looking at specimens very closely. Most laboratory microscopes give a magnification up to about 200 times but some can give a magnification of over 1000 times. The microscope must also provide a clear view and this is achieved by controlling the amount of light shining onto the specimen.

Light is collected by a mirror at the base of the microscope. The mirror is attached so that it can move in any direction. The light comes from a lamp or from a sunless sky. It must never be collected directly from the Sun as this can cause severe eye damage and blindness. Some microscopes have a built-in lamp instead of a mirror. The light either shines directly through a hole in the stage onto the specimen or it passes through a hole in a diaphragm. The diaphragm allows the amount of light reaching the specimen to be controlled by increasing or decreasing the size of the hole. The light shining through a specimen is called transmitted light.

Above the specimen is the ocular tube. This has an eyepiece lens at the top and one or more objective lenses at the bottom. The magnification of the two lenses is written on them. An eyepiece lens may give a magnification of ×5 or ×10. An objective lens may give a magnification of ×10, ×15 or ×20. The magnification provided by both the eyepiece lens and the objective lens is found by multiplying their magnifying powers together. Most microscopes have three objective lenses on a nosepiece at the bottom of the ocular tube. The nosepiece can be rotated to bring each objective lens under the ocular tube in turn.

An investigation with the microscope always starts by using the lowest power objective lens then working up to the highest power objective lens if it is required.

A specimen for viewing under the microscope must be put on a glass slide. The slide is put on the stage and held in place by the stage clips. The slide should be positioned so that the specimen is in the centre of the hole in the stage.

The view of the specimen is brought into focus by turning the focusing knob on the side of the microscope. This may raise or lower the ocular tube or it may raise or lower the stage on which the slide of the specimen is held. In either case you should watch from the side of the microscope as you turn the knob to bring the objective lens and specimen close together. If you looked down the ocular tube as you did this you might crash the objective lens into the specimen, which could damage both the lens and the specimen. When the objective lens and the specimen are close together, but not touching, look down the eyepiece and turn the focusing knob so that the objective lens and specimen move apart. If you do this slowly, the blurred image will become clear.

11 What is a microscope used for?

12 What advice would you give someone about how to collect light to shine into a microscope?

13 What magnification would you get by using an eyepiece of ×5 magnification with an objective lens of ×10 magnification?

14 If you had a microscope with ×5 and ×10 eyepieces and objective lenses of ×10, ×15 and ×20, what powers of magnification could your microscope provide?

15 How would you advise someone to use the three objective lenses on the nosepiece?

16 Why should you not look down the microscope all the time as you try to focus the specimen?

17 Look at the picture of the microscope in Figure 1.4 and describe the path taken by light from a lamp near the microscope to the eye.

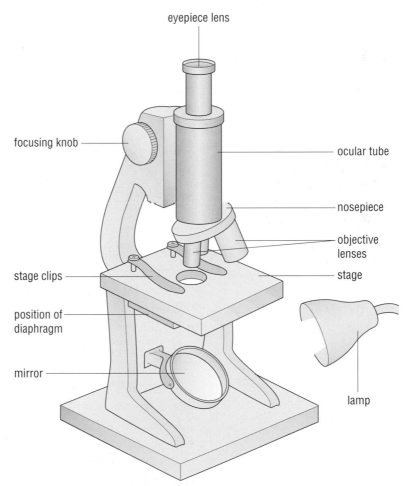

Figure 1.4   The main parts of a microscope.

Finding the size of microscopic specimens

The disc of light you see when you look down a microscope is called the field of view. You can estimate the size of the specimens you see under the microscope if you know the size of the field of view. A simple way to find the size of the field of view is to put a piece of graph paper on a

18 A field of view was found to be 2000 μm in diameter. A soil particle reached one quarter of the way across it. How long would you estimate the length of the particle to be?

19 The fields of view of three lenses were measured. A was 100 μm, B was 3000 μm and C was 500 μm. Which was the most powerful lens and which was the least powerful lens?

20 Why does the field of view decrease as the power of the objective lens increases?

slide and examine it using the low power objective lens. The squares on the graph should be 1 mm across. Microscopic measurements are not made in millimetres, but in micrometres. 1 mm = 1000 micrometres (written as 1000 μm).

If the field of view is two squares across it has a diameter of 2000 μm. If you remove the slide with the graph paper and replace it with a slide with some soil particles, you could estimate the size of a soil particle by judging how far it crosses the field of view.

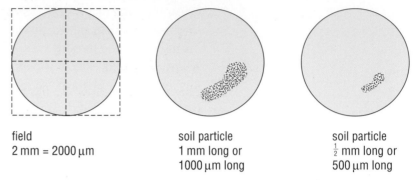

field
2 mm = 2000 μm

soil particle
1 mm long or
1000 μm long

soil particle
$\frac{1}{2}$ mm long or
500 μm long

Figure 1.5 A soil particle under the microscope.

If the soil particle comes halfway across the field of view it is 1000 μm long. There is a relationship between the power of an objective lens and its field of view. As the power of the objective lens increases, the size of its field of view decreases.

Cells

There are ten times more cells in your body than there are people on the Earth. If you stay in the water a long time at a swimming pool you may notice that part of your skin sometimes flakes off when you dry yourself. These flakes are made of dead skin cells.

Figure 1.6 shows a section of human skin that has been stained and photographed down a microscope using a high power objective lens. When unstained, the different parts of the cells are colourless and are difficult to distinguish. In the 1870s it was discovered that dyes could be made from coal tar which would stain different parts of the cell. Cell biologists found they could stain the nucleus and other parts of the cell different colours to see them more easily.

21 Why are most specimens of cells stained before they are examined under the microscope?

22 You look down a microscope at a slide labelled 'Cells'. You can see a coloured substance with dots in it and lines that divide the substance into rectangular shapes. Inside the rectangular shapes, what are:
   a) the dots,
   b) the lines,
   c) the coloured substance?

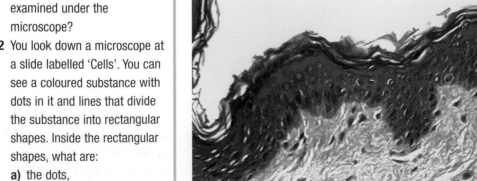

Figure 1.6 Section of human skin. Cells can be seen flaking off the surface.

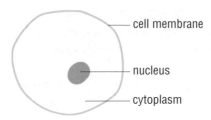

Figure 1.7 A typical animal cell.

23 If there are about 6000 million people on the Earth, how many cells have you got in your body?
24 How does the cell membrane protect the cell?

Basic parts of a cell

Nucleus

This is the control centre of the cell. It contains the genetic material, called DNA (its full name is deoxyribonucleic acid). The DNA molecule is a long chain of smaller subunits. They occur in different combinations along the DNA molecule. The combinations of subunits provide instructions for the cell to make chemicals to keep it alive or to build its cell parts. When a cell divides the DNA divides too, so that the nucleus of each new cell receives all the instructions to keep the new cell alive and enable it to grow.

Cytoplasm

This is a watery jelly which fills most of the cell in animal cells. It can move around inside the cell. The cytoplasm may contain stored food in the form of grains. Most of the chemical reactions that keep the cell alive take place in the cytoplasm.

Cell membrane

This covers the outside of the cell and has tiny holes in it called pores that control the movement of chemicals in or out of the cell. Dissolved substances such as food, oxygen and carbon dioxide can pass through the cell membrane. Some harmful chemicals are stopped from entering the cell by the membrane.

Additional parts found only in plant cells

Cell wall

This is found outside the membrane of a plant cell. It is made of cellulose, which is a tough material that gives support to the cell.

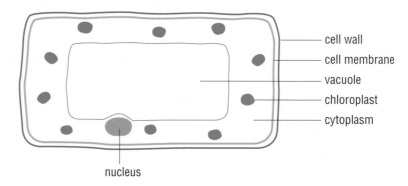

Figure 1.8 A typical plant cell.

Chloroplasts

These are found in the cytoplasm of many plant cells. They contain a green pigment called chlorophyll which traps a small amount of the energy in sunlight. This energy is used by the plant to make food in a process called photosynthesis (see Chapter 11). Chloroplasts are found in many leaf cells and in the stem cells of some plants.

25 Name two things that give
support to a plant cell.
26 Would you expect to find
chloroplasts in a root cell?
Explain your answer.
27 Why do plants wilt if they are
not watered regularly?

Large vacuole

This large space in the cytoplasm of a plant cell is filled with a liquid called cell sap which contains dissolved sugars and salts. When the vacuole is full of cell sap the liquid pushes outwards on the cell wall and gives it support. If the plant is short of water, the support is lost and the plant wilts.

Some animal cells and Protoctista (see page 65) have vacuoles but they are much smaller than those found in plant cells.

Adaptation in cells

The word adaptation means the change of an existing design for a particular task. The basic designs of animal and plant cells were shown in the last section, but many cells are adapted which allows them to perform a more specific task. Here are some common examples of the different types of animal and plant cells.

Ciliated epithelial cells

Cells that line the surface of structures are called epithelial cells. Cilia are microscopic hair-like extensions of the cytoplasm. If cells have one surface covered in cilia they are described as ciliated. Ciliated epithelial cells line the throat, for example. Air entering the throat contains dust that becomes trapped in the mucus of the throat lining. The cilia wave to and fro and carry the dust trapped in the mucus away from the lungs.

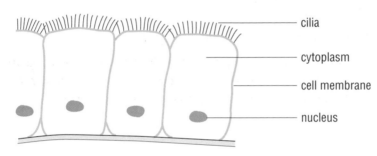

Figure 1.9 Ciliated epithelial cells.

Sperm cells

These transport the male genetic material in their nucleus. They have a streamlined shape which allows them to move easily through the liquid as they travel towards the female egg (see Figure 1.10). They have a tail that waves from side to side to push the cell forwards.

Egg cells

These contain the female genetic material in their nucleus. They are much larger than sperm cells because they contain a food store, and they do not move on their own. When a sperm reaches an egg the genetic material from the sperm combines with that in the egg in a

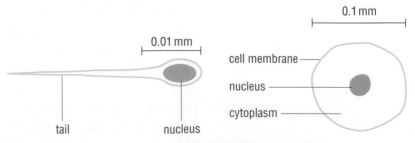

Figure 1.10 Sperm (left) and egg cell (right). Note size scales.

process called fertilisation. This produces a cell which divides many times to produce an embryo. The food and energy for early growth of the embryo is provided by the egg.

Root hair cells

These plant cells grow a short distance behind the root tip. The cells have long thin extensions that allow them to grow easily between the soil particles. The shape of these extensions gives the root hair cells a large surface area through which water can be taken up from the soil.

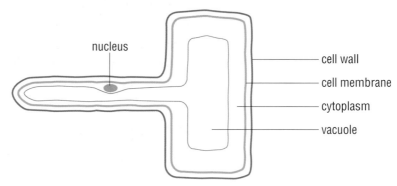

Figure 1.11   A root hair cell.

Leaf palisade cells

These plant cells have a shape that allows them to pack closely together in the upper part of a leaf, near the light. They have large numbers of chloroplasts in them to trap as much light energy as possible.

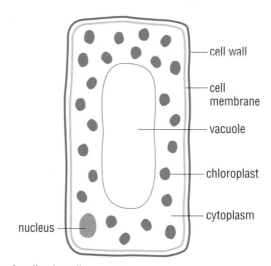

Figure 1.12   A palisade cell.

28 How are sperm and egg cells
   a) similar,
   b) different?
29 Smoking damages the cilia lining the breathing tubes. What effect might this have on breathing?
30 What changes have taken place in the basic plant cell to produce a root hair cell?
31 How is a palisade cell different from a root hair cell? Explain these differences.
32 Why would it be a problem if root hair cell extensions were short and stubby?
33 Why are there different kinds of cells?

From cells to tissues

Cells are not mixed up inside the body. They are arranged in an orderly way in groups. A group of cells of the same type is called a tissue, and it performs an important task in the life of an organism. Figure 1.6 on page 6 shows a tissue of cells. They are called epithelial cells and their task is to make a waterproof and germ-proof covering over the body. Figure 1.9 on page 8 shows some cells from a tissue in the throat. The function of this tissue is to transport harmful particles away from the lungs.

Tissues in plants

If you examine a very thin slice of a leaf (called a section) under the microscope you can see how it is composed of different tissues of cells (see Figure 1.13).

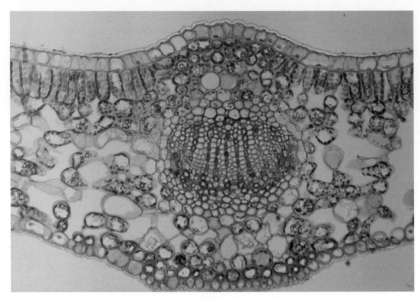

Figure 1.13 Section through a plant leaf.

34 What is a tissue?

35 a) Name the tissues labelled A, B and C in the section of this leaf.

   b) What is X?

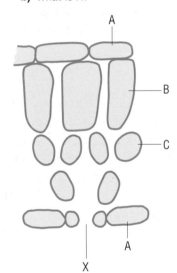

Figure 1.14

36 a) Compare the tissue of the epidermis and the palisade tissue of the leaf.

   b) How are the tasks of the palisade tissue and the spongy mesophyll

   i) similar,

   ii) different?

Covering tissue

The surface of the leaf is covered with a tissue of flat, transparent cells which make up the epidermis. This surface allows light rays to pass through the tissues of cells inside the leaf. It also contains holes called stomata which regulate water loss and gas exchange.

Energy-trapping tissue

Under the upper epidermis is the palisade tissue, made from tightly packed palisade cells (see Figure 1.12, page 9). These cells contain many chloroplasts and trap some of the energy in sunlight to make food (see Chapter 11).

Spongy tissue

Below the palisade tissue is a tissue called the spongy mesophyll. The cells in this tissue can also collect energy in sunlight and make food but their main purpose is to provide a large surface area inside the leaf through which water can evaporate and gases can be exchanged.

Conducting tissue

Water moves through the plant from the roots to the leaves in xylem tissue. This tissue is made from cells which die and form hollow tubes called xylem vessels. Water is needed in the leaf not only to keep the leaf cool, but also as a raw material for making food.

There is a second tissue which forms tubes – this is called phloem. The tubes are made of living cells which transport the food made in the leaf to other parts of the plant.

From tissues to organs

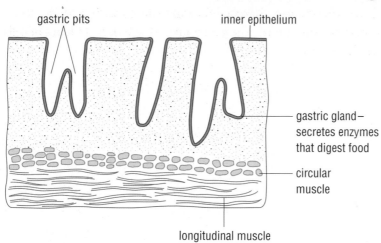

gastric pits

inner epithelium

gastric gland – secretes enzymes that digest food

circular muscle

longitudinal muscle

Figure 1.15   Section through stomach wall.

Just as cells are arranged in an orderly way into tissues, tissues are arranged in an orderly way into groups called organs. An organ may perform one or more tasks to help keep an organism alive, and organs can contain different types of tissues.

The human stomach wall is an organ made from several tissues. These include muscle tissues which help churn up the food in the stomach, and tissues which form glands that produce food-digesting liquids.

Cells and reproduction

Most living things are produced by the joining of two special cells called gametes. Gametes are the reproductive cells of an organism.

In animals

In animals, the male produces a gamete called the sperm cell and the female produces a gamete called the egg cell (see Figure 1.10, page 8). These two gametes join together in a process called fertilisation (see Chapter 2).

In flowering plants

Plants also produce male and female gametes. The male gamete is a cell in the pollen grain and the female gamete is an egg cell in the ovule. Most flowers have both male and female parts, as Figure 1.16 shows.

The transfer of the male gametes from the anthers to the stigma of the female carpel is called pollination. If the transfer occurs within a single plant it is called self-pollination; if it is between different plants it is called cross-pollination. Transfer is usually by wind or insects.

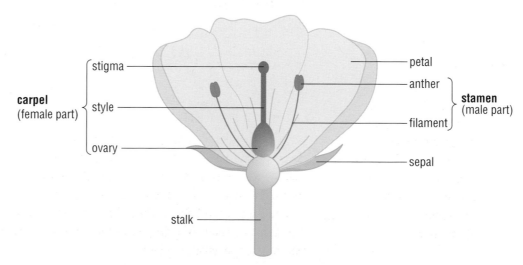

stigma

petal

anther

carpel (female part)

style

stamen (male part)

filament

ovary

sepal

stalk

Figure 1.16   Parts of a typical insect-pollinated flower.

Fertilisation

After a pollen grain has reached the surface of a stigma it breaks open and forms a pollen tube. The male gamete that has travelled in the pollen grain moves down this tube. The pollen tube grows down through the stigma and the style into the ovary (see Figure 1.17). In the ovary are ovules, each containing a female gamete. When the tip of a pollen tube reaches an ovule the male gamete enters the ovule. It fuses with the female gamete in a process called fertilisation and a zygote is produced.

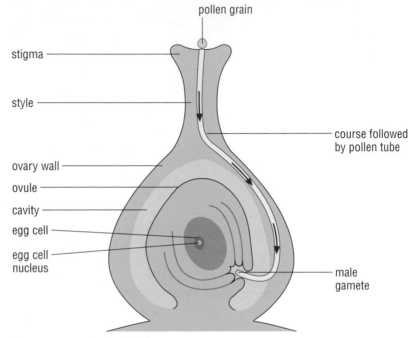

Figure 1.17 Fertilisation.

37 What is the difference between pollination and fertilisation?

38 Trace the path of a male gamete nucleus from the time it forms in a pollen grain in an anther until the time it enters an ovule.

After fertilisation

The zygote undergoes repeated cell division to form the embryo plant. Structures that later become the root and shoot are developed and a food store is laid down. While these changes are taking place inside the ovule the outer part of the ovule is forming a tough coat. When the changes are complete the ovule has become a seed (see Figure 1.18). As the seeds are forming other changes are taking place. The petals and

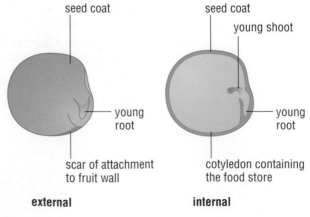

Figure 1.18 Parts of a pea seed.

stamens fall away. The sepals usually fall away too but sometimes, as in the tomato plant, they may stay in place. The stigma and style wither and the ovary changes into a fruit.

Types of fruit

A fruit forms from the parts of the flower that continue to grow after fertilisation. There are two main types of fruit – dry fruits and succulent fruits.

Dry fruits have a wide variety of forms. They may form pods, such as those holding peas and beans, they may be woody nuts, such as acorns or hazelnuts, or grains like the fruits of wheat, oats and grasses.

Succulent fruits have a soft fleshy part. They may have a seed inside a woody skin which forms a 'stone' in the fruit, as in the cherry and peach. Many succulent fruits do not have a stone but contain a large number of smaller seeds, as in the tomato and orange.

Some fruits, such as apples, are called false fruits because their fleshy part does not grow from part of the flower but from the receptacle on which the flower grows.

Summary

- The part of a body that performs a task to keep a living thing alive is called an organ (*see page 2*).
- There are ten organ systems in the human body. They are the sensory system, the nervous system, the respiratory system, the digestive system, the circulatory system, the excretory system, the skeletal system, the muscle system, the endocrine system and the reproductive system (*see pages 2–4*).
- There are four main organs in the body of a flowering plant. They are the root, stem, flower and leaf (*see page 4*).
- The microscope is used to observe very small living things or the cells of larger living things (*see page 4*).
- There are special techniques for finding the size of a small object with a microscope (*see page 5*).
- The bodies of plants and animals are made of cells. The basic parts of the cell are the nucleus, cytoplasm and cell membrane (*see pages 6–7*).
- In a plant cell there is a cellulose cell wall and a vacuole (*see pages 7–8*).
- Cells have different forms for different functions. They are adapted to perform specific tasks in the body and life of the organism (*see page 8*).
- Cells form groups called tissues (*see page 9*).
- Plants and animals produce sex cells called male and female gametes (*see page 11*).
- The parts of a flower are adapted for reproduction (*see page 11*).
- Pollen is transferred in pollination (*see page 11*).
- When fertilisation takes place in flowering plants, seeds and fruits are produced (*see page 12*).

1 If organisms as complex as us have developed on a planet similar to Earth but with a stronger force of gravity, what body features would you expect them to have?

2 Pollen grains will grow pollen tubes if they are placed in a sugar solution of a certain concentration. The results in the tables reflect results sometimes produced in investigations.

An experiment was set up to find out the concentration of sugar that would cause the pollen grains of a plant to produce pollen tubes. Here is a table of the results.

Table 1.1

Concentration of sugar solution (%)	5	10	15	20
Number of pollen grains	20	20	20	20
Number of grains with tubes	4	18	3	0

a) What percentage of pollen grains produced tubes in each solution?

b) What would you expect the concentration of sugar in the stigmas of the flowers to be?

When the pollen of a second type of plant was investigated the following results were obtained.

Table 1.2

Concentration of sugar solution (%)	5	10	15	20
Number of pollen grains	20	20	20	20
Number of grains with tubes	0	1	3	8

c) Why was it decided to take the investigation further?

d) What do you think was done to take the investigation further?

Reproduction

All animals follow a particular pattern of development. Like plants, they grow from a fertilised egg and after hatching or being born they grow until they are fully mature. At this stage they are capable of reproduction. In almost all kinds of animal there are those individuals that produce eggs – the females, and those that produce sperm – the males. For reproduction to occur the sperm from the male has to reach the eggs of the female, and the head of one sperm has to enter an egg. When the sperm and egg nuclei fuse, fertilisation has occurred.

Reproduction in humans

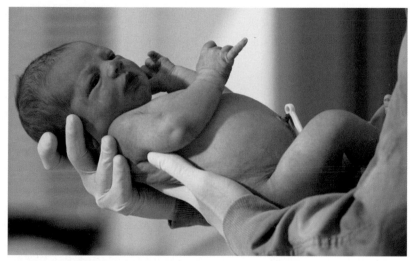

Figure 2.1 This new human was born only seconds ago.

How the changes begin

The bodies of newborn male and female babies are very similar. The sex of a baby can only be declared after looking at the baby's genitalia – the external parts of the reproductive organs. For the next 10 years the bodies of boys and girls continue to be very similar (except for the genitalia), then changes begin.

Behind the nose and beneath the brain is an organ called the pituitary gland. This is one of several glands that secrete different hormones into the blood. A hormone is a chemical that is produced in one part of the body, circulates all round the body in the blood but only has an effect on a specific part of the body. At around the age of 10 to 13 in girls and 12 to 14 in boys, the pituitary gland secretes increased amounts of hormones controlling growth and sexual development.

Growth hormone travels around the body and stimulates growth in the hands and feet, then the hips and chest, then causes the trunk followed by the legs to increase in length. These changes in body size can make people feel uncomfortable.

In females, the sex hormones produced in the pituitary gland control the development of the ovaries. In turn, the ovaries produce two sex hormones, oestrogen and progesterone. Cyclical fluctuations in the pituitary and ovarian hormones result in the monthly release of eggs and the onset of the menstrual cycle. In males, the sex hormone testosterone is produced in the testes. The sex hormones not only cause the reproductive organs to develop fully but also cause the development of body features known as secondary sexual characteristics. The sex hormones work with the growth hormone. In females they cause the hips to grow wider than the shoulders and to develop fat around them to give a smoother shape. In males, testosterone causes the shoulders to grow wider and heavier than the hips.

Table 2.1 Secondary sexual characteristics.

Males	Females
Growth of hair on the face, armpits and pubic region	Growth of hair in armpits and pubic region
Voice becomes deeper	Breasts develop
Growth of penis and testicles	Growth of vagina and uterus
	Pelvis widens

Table 2.1 shows the secondary sexual characteristics that develop in males and females due to the action of these sex hormones. These changes take place over a period of about 3 years in girls but may take longer in boys. There is great variation in the time the changes begin and also in the nature and the size of the changes. These variations can make people unnecessarily anxious and cause them to worry if they feel they are not changing in the correct way. When the changes are complete each person is capable of reproduction. This period of change is called puberty. It takes place in the first half of adolescence, which is the time when a person changes from a child to an adult. In adolescence, in addition to becoming sexually mature, a person also develops adult emotions and social skills.

Figure 2.2 Most of the people in this school hall are going through puberty, while the rest went through puberty many years ago.

1 How may the body of a girl and a boy change from the beginning to the end of puberty?
2 Why do some people feel anxious as they go through puberty?

All mixed up

The greatest physical and emotional changes in a person's life take place during adolescence. At the end of this time people become adults, usually with a wider range of social skills to allow them to live independently, if they wish, and to develop their own ambitions for a happy life.

During adolescence people have to learn to cope with both the physical and emotional changes at once. In addition to the physical changes, girls have to learn to cope with periods, and boys may have 'wet dreams' in which they release semen when they are asleep. A person's main emotional change is in wanting to be more independent and have more control over their life. Generally, other members of the family adjust to these changes and the adolescent is allowed more freedom in choice of clothes and use of free time. Many schools also provide a wide range of sports, clubs and activities to help the adolescent to develop emotionally and socially.

It is natural for people in adolescence to be anxious sometimes and to worry whether they are going through the proper physical change. Some also try to be too independent too quickly. This may lead to arguments with parents and other adults. As well as seeking independence from their families they are also anxious not to lose their friends and may feel under pressure to follow the way their friends behave, even though they would prefer to do something else. The sex hormones also influence thoughts and make people interested in the opposite sex. The degree to which the sex hormones do this varies as much as the physical changes they make. All the changes taking place over a few short years sometimes make many adolescents feel confused. This is natural. Some people find that talking to others helps them cope, while other people feel embarrassed to talk about how they feel.

For discussion

Is this a fair description of how people think and feel during adolescence?

What would you add to this account and what do you not agree with?

Male reproductive organs

The testes develop inside the body of the unborn child and usually move down to the outside of the body before the baby is born. The testes are held in a bag of muscle and skin called the scrotum. They are positioned outside the body because the conditions are cooler there and are more favourable for sperm production. Each testis contains long microscopic pipes called seminiferous tubules. The male gametes or sex cells are made inside these tubes. They are the sperm cells. On the top and side of each testis is an epididymis. This is a long, coiled tube in which the sperm cells collect. The sperm cells travel to the outside of the body along the sperm duct and the urethra. There are glands along the

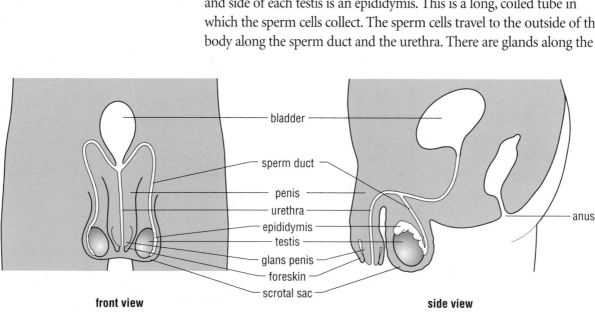

front view

side view

Figure 2.3 The male reproductive system.

path to the outside that add liquid to the sperm cells. The mixture of liquid is known as seminal fluid, or semen.

The urethra runs through the middle of the penis. It is also connected to the bladder and is the tube through which the urine flows. Semen and urine do not flow down the urethra at the same time.

The penis contains spongy tissue along its length that can fill with blood to make it hard, stiff and erect. The tip of the penis, called the glans penis, has a large number of receptors and is very sensitive. The glans is covered with a fold of skin called the foreskin. If circumcision has taken place the foreskin will have been removed.

Female reproductive organs

The ovaries develop inside the body and, unlike the testes, they stay there because egg production can take place at body temperature. Each ovary contains about 200 000 potential egg cells. Egg cells are the female gametes. Eggs are released as part of the menstrual cycle (see below). When an egg is released from an ovary it passes into the trumpet-shaped opening of the oviduct – a tube that connects to the uterus or womb. If an egg is fertilised (see page 21) it develops into a fetus in the uterus.

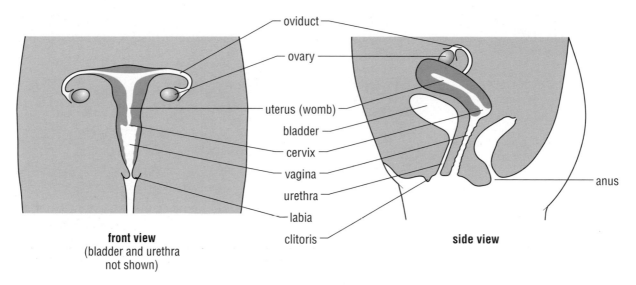

front view
(bladder and urethra
not shown)

oviduct
ovary
uterus (womb)
bladder
cervix
vagina
urethra
labia
clitoris

side view

anus

Figure 2.4   The female reproductive system.

3 How are the male and female reproductive organs different?
4 How are they similar?

The uterus is connected to the vagina by the cervix. The vagina opens to the outside just behind the opening of the urethra. Both openings are protected by folds of skin called labia. These folds also protect a region about the size of a small pea called the clitoris. This region has a large number of receptors like the tip of the penis.

Menstrual cycle

From the beginning of puberty the menstrual cycle occurs approximately every month in females. It does not take place when the female is pregnant. It includes a period of bleeding from the vagina which may last for about 4 days. During this time an egg starts to mature in one of the ovaries. About 10 days after the period of bleeding ends, the egg, which is about the size of the dot on this i, escapes from the ovary. Alternate ovaries release an egg each month. The egg is then moved down the fluid-filled oviduct by the movement of cilia in the oviduct walls.

At the same time as the egg is maturing in the ovary, the uterus wall is thickening with blood. It does this to prepare to receive a newly formed embryo in case fertilisation takes place. The egg may survive for up to 2 days in the oviduct and fertilisation can take place during this time. If the egg is not fertilised no further development of the egg takes place. About 12 days after the egg dies, the uterus wall breaks down and blood passes out of the vagina. Another menstrual period begins.

The length of time of the period of bleeding varies between girls and so does the amount of blood that is released. Some girls and women feel ill a day or two before their period starts or feel pain for the first few days, while others are not affected in this way.

The menstrual cycle continues until the beginning of the menopause, which may start at about the age of 45. During the menopause periods may become irregular and eventually stop. The menopause usually ends when a woman is in her early 50s.

5 How long is the average menstrual cycle?

6 How does the wall of the uterus change during the course of the cycle and why?

7 What is the cause of the menstrual bleeding?

8 In what ways can periods vary?

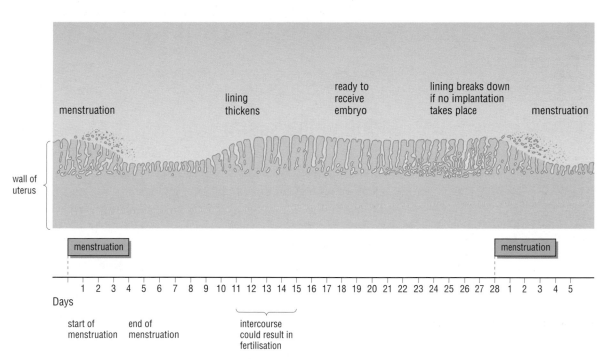

Figure 2.5 Changes in the wall of the uterus during the menstrual cycle.

Sexual intercourse

Before sexual intercourse can take place the penis must become erect. This happens by the action of a muscle at the base of the penis. It prevents the drainage of blood from the penis. The blood collects in the spongy tissue and makes it expand and become hard.

Prior to intercourse the vagina may also widen to ease the passage of the penis into it. The lining of the vagina may secrete a fluid that acts as a lubricant and further helps the penis to enter the vagina.

When the penis is inside the vagina both the male and female may make thrusting movements to stimulate the sensitive areas of the penis tip and the clitoris. This may give each partner a feeling of pleasure called an orgasm. When the male has an orgasm it is accompanied by a contraction of the muscles in the epididymis and sperm ducts which propels the semen through the penis into the vagina. The action of releasing the semen is called an ejaculation. The volume of semen ejaculated is usually about 3–5 cm^3.

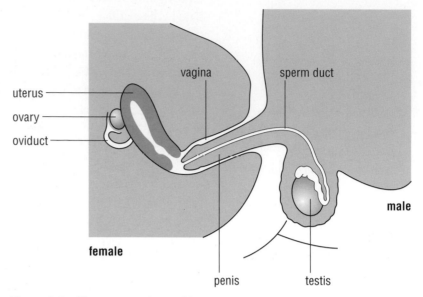

Figure 2.6   The process of sexual intercourse.

Path of the sperm

The semen ejaculated into the top of the vagina may contain over 400 million sperm. The sperm cells do not contain a food store to provide them with energy for movement. They get their nourishment from the seminal fluid which was secreted by glands as the sperm made their way from the testicles to the urethra. The food provides each sperm with energy to lash its tail like a whip. This movement drives the sperm forwards.

The sperm travel through the cervix and up the mucus lining of the uterus wall into the oviducts. Millions die on the way leaving only a few thousand to enter the oviduct. As the sperm swim along the oviduct even more die so that only a few hundred reach their destination 4–6 hours after ejaculation. The sperm may survive for 2 or 3 days here before they die. During this time, if the sperm meet an egg, fertilisation may occur.

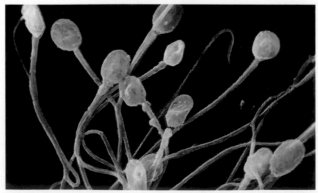

Figure 2.7   Photomicrograph of sperm cells.

How the egg is moved

The egg is much larger than a sperm cell because it contains its own food store. This is to provide energy and materials for the very early development of the embryo if the egg is fertilised. Unlike the sperm, the egg has no means of propulsion. It is moved by the action of the cilia in the wall of the oviduct. They wave backwards and forwards and push the egg along.

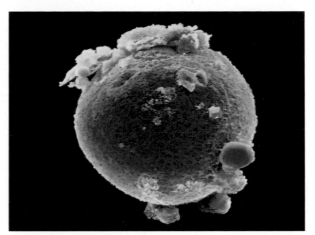

Figure 2.8   Photomicrograph of an egg.

9 Compare the ways in which the sperm and egg move along the oviduct.

Fertilisation

When the sperm cells meet an egg in the oviduct they crowd around it. The head of only one sperm cell penetrates the cell membrane of the egg (see Figure 2.9). This sperm cell's head breaks off from the tail and moves through the egg cell's cytoplasm to the nucleus. When the head reaches the egg cell's nucleus fertilisation takes place. In this process the nucleus inside the sperm head fuses with the egg cell nucleus. The fertilised egg is called a zygote. Changes to the cell membrane around the zygote prevent other sperm heads from entering and fusing with the nucleus.

The nuclei of the sperm and the egg contain instructions for the development of the baby. You can read about these on page 86.

If the fertilised egg becomes implanted into the uterus wall, conception has taken place. It results in pregnancy.

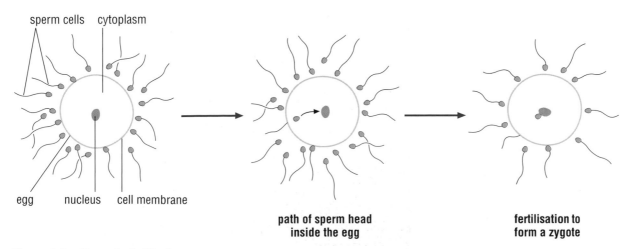

sperm cells   cytoplasm

egg   nucleus   cell membrane

path of sperm head
inside the egg

fertilisation to
form a zygote

Figure 2.9   Stages in fertilisation.

Development of the baby

After fertilisation the zygote divides into two cells, then four, then eight and so on (see Figure 2.10). The zygote does not increase in size in the first 7 days, so cells become smaller at each division. By 7 days after fertilisation the cells have formed a hollow ball and have reached the uterus. The hollow ball sinks into the thick lining of the uterus wall, which has a large amount of blood passing through it. This process of sinking into the uterus wall is called implantation.

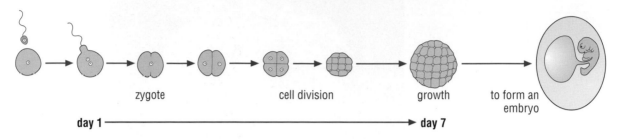

zygote   cell division   growth   to form an embryo

day 1 ⟶ day 7

Figure 2.10 The early development of the zygote into an embryo.

Placenta

The cells on the surface of the hollow ball of cells of the zygote will form the placenta. This takes in food and oxygen from the mother's blood so that the cells can grow and divide. Waste products, such as carbon dioxide from the cells' activities, pass across the placenta to the mother's blood so she can remove them through her own lungs or kidneys.

During the course of the pregnancy the placenta grows into a disc with a diameter of about 20 cm. It forms microscopic finger-like projections called villi that penetrate into the uterus wall and make a very large surface area for the exchange of materials between the mother and her baby. The placenta is attached to the developing baby by the umbilical cord. Blood runs through vessels in this cord between the placenta and the baby's tissues. The baby's blood and the mother's blood always remain separate. The mother's blood is at a much higher pressure than the baby's blood so it would damage its blood vessels if it passed directly through the placenta. The two kinds of blood may not be compatible. This means that if they mixed, clotting would occur which would block the blood vessels and lead to further damage.

The placenta makes hormones that stop the ovaries producing any more eggs, and which keep the uterus wall from breaking down as it would normally do in the menstrual cycle. Antibodies pass from the mother's blood through the placenta to her baby to give protection from diseases.

Embryo and fetus

Although most people refer to a baby growing in the uterus, the words embryo and fetus are often used for the early stages. While some of the cells inside the hollow ball join with those on the outside to make the placenta, most of the cells form the embryo. At the end of the first 2 weeks of development the embryo is a flat disc of tissue, but by 4 weeks it has developed a simple body shape with stumps where the limbs will

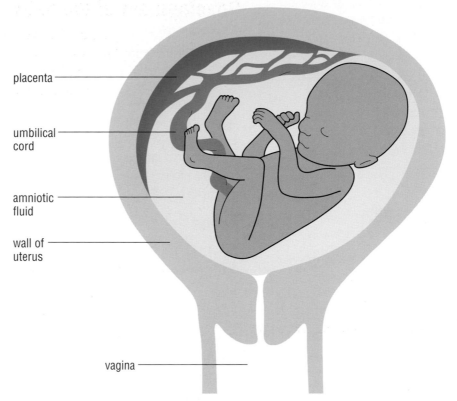

placenta

umbilical cord

amniotic fluid

wall of uterus

vagina

Figure 2.11 The fetus in the uterus

grow. Internally its heart has started to beat. By 8 weeks all of the organ systems have formed and the embryo, which is now 2.5 cm long, is called a fetus.

The fetus continues to increase in size and the organ systems become more fully developed. By 14 weeks the sex of the fetus can be revealed by an ultrasound scan. By 20 weeks the fetus is 12.5 cm long but its legs are growing quickly. Eight weeks later the fetus turns upside down with its head towards the cervix. At the end of 38 weeks after fertilisation of the egg the fetus is about 50 cm long and has a mass of about 3 kg.

The baby is now ready to be born. The length of time between fertilisation and birth is called the gestation period. In medicine this is referred to as 40 weeks (rather than 38 weeks) because the time is calculated from the first day of the last menstrual period.

Amnion and its fluid

During the development of the fetus a bag forms around it called the amnion, which contains watery fluid. The fluid acts like a cushion around the fetus and protects it from pressures outside the uterus that might squash it. The fluid also allows the fetus to float freely so that growing limbs have space to develop and are not pressed against the wall of the uterus where their growth would be restricted.

Damaging the fetus

If the mother smokes during her pregnancy the amount of oxygen reaching the fetus is reduced. This may hinder the baby's development and might result in the baby's mass being as much as 200 g below normal when it is born.

10 How do the placenta and the amniotic fluid help the embryo and fetus develop?

11 How is an embryo different from a fetus?

12 How is the fetus put at risk if the mother smokes or drinks alcohol while she is pregnant?

Drinking alcohol regularly during pregnancy increases the risk of miscarriage. This occurs when the placenta detaches from the uterus wall and the fetus dies. Drinking alcohol also increases the chance of the newborn baby being less well developed mentally and physically.

Drugs such as LSD and amphetamines contain poisons that affect the healthy development of the fetus. If the mother is a heroin addict the baby will also be addicted.

Rubella is a virus that causes a relatively mild illness in adults, but which is very harmful to fetuses. The virus can pass across the placenta from the mother and damage the fetus during the first three months of pregnancy.

Twins

About one in a hundred pregnancies produces twins. The twins may be identical or non-identical. Identical twins form from the same fertilised egg. At the first cell division of the zygote (see Figure 2.10, page 22) the two cells move apart and each one develops into an embryo. Non-identical twins form from two eggs that are released into the oviduct at the same time. Each egg is fertilised by a different sperm. Identical twins share the same genetic material but non-identical twins have different genetic material.

In some very rare circumstances, conjoined (popularly called Siamese) twins develop. They are identical twins but the cells from which they formed did not separate completely. They may be joined at the head, the hip or the chest. They may have two sets of organs, or one twin may be without one or more organs and rely on the other for survival.

13 A boy and girl are a pair of twins. Are they identical or non-identical? Explain your answer.

14 What could be the consequence of separating conjoined twins who share some organs? Explain your answer.

Figure 2.12

Identical twins

Non-identical twins

Birth

There are three stages to the birth process. In the first stage the muscles in the uterus wall begin to contract. The time between each contraction may be up to 30 minutes at first. During the 12 to 14 hours of this first stage the time between contractions shortens to 2 to 3 minutes. At some point the contractions cause the amnion to break and the fluid to pass out down the vagina. This is called the 'breaking of the waters'. At the end of the first stage the cervix has widened so that the head of the fetus can start to pass down it.

In the second stage the mother contracts her abdominal muscles with the contractions of the uterus to push the fetus down the birth canal. This stage may only last a few minutes and is completed when the baby has been born and the umbilical cord has been cut and clamped to prevent loss of blood. If there are twins the contractions stop for about 10 minutes after the first baby has been born then start again. Only a few contractions may be needed for the second baby to be born.

The two stages from the first contractions of the uterus to the birth of the baby are called labour.

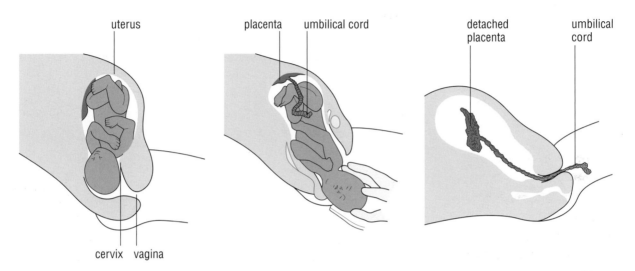

Figure 2.13 Diagrammatic representation of birth.

The third stage lasts for about 20 minutes after the second. During this time the placenta comes away from the wall of the uterus and passes down the vagina. When it has left the mother the placenta is called the afterbirth.

Summary

- All animals follow a particular pattern of development (*see page 15*).
- Changes in the human body at puberty are brought about by the sex hormones (*see pages 15–16*).
- The male and female reproductive organs have differences and similarities (*see pages 17 and 18*).
- The menstrual cycle occurs due to the monthly release of an egg (*see pages 18–19*).
- Fertilisation is the fusion of a sperm nucleus with an egg nucleus (*see page 21*).
- Development of the embryo takes place in the uterus (*see pages 22–23*).
- The placenta and the amnion play important parts in the development of the embryo and fetus (*see pages 22 and 23*).
- Twins may be identical or non-identical (*see page 24*).
- The uterus and abdominal muscles are used in the birth process (*see above*).

1 How does a knowledge of the basic facts of reproduction help someone going through puberty?

2 Gillian and Barry have got a record of their heights (in cm), measured once a year from birth until they were 17 years old.

Gillian's record is: 51, 75, 87, 95, 102, 108, 115, 122, 127, 132, 138, 144, 151, 157, 159, 161, 162, 163

Barry's record is: 51, 76, 88, 96, 103, 109, 117, 123, 129, 135, 139, 143, 149, 154, 160, 167, 173, 174

a) Make a table of the results.

b) Plot both sets of results on the same graph.

c) At what ages were Gillian and Barry the same height?

d) Who grew more quickly between the ages of
  i) 8 and 9,
  ii) 11 and 13,
  iii) 14 and 17?

For discussion

'People should develop a responsible attitude towards sex.' What do you think this means?

3 *Survival*

If you look out across the countryside you may see fields, hedges, woods, ponds and maybe a river. Most of the living things you see will be plants ranging in size from green slime on rocks to the tallest tree in a wood. You may see some birds flying across the countryside and a few insects moving through the air around you. There may be a slug slowly moving across your path or a squirrel scampering away through the branches of a nearby tree. The scene may look too complicated to investigate scientifically, but the study of ecology was established at the beginning of the 20th century to do just this.

Ecology means the study of living things and where they live. The home area of a living thing is called its habitat. Two examples of habitats are a wood and a pond. The country scene in Figure 3.1 can be divided into a number of different habitats for further investigation.

A living thing in its habitat is affected by two different kinds of factors. They are abiotic factors and biotic factors. Abiotic factors are not due to living things and include temperature, wind strength, amount of light and moisture. Biotic factors are due to living things and include plants and animals as sources of food, other organisms competing for space, and predators.

> **1** What habitats can you see in Figure 3.1?

Adaptation

A living thing survives in its habitat because it can cope with the abiotic and biotic factors. It copes if its body is adapted to the conditions of the habitat. Changes that have taken place in the structures of different species over time that help them survive are called adaptations.

Figure 3.1 A countryside scene from a hilltop showing a range of habitats.

Some adaptations in plants

Grass

2 Why do the low growing points of grasses help them to survive?

One of the commonest land plants is grass. Unlike many plants it survives in areas where there are grazing animals. Grass has growing points that are below the reach of the grazers' mouths. This means that when most of the leaves of the grass are removed and eaten, the growing points can produce new leaves. The grazing animals also eat the grass flowers, but grass can also reproduce asexually by sending out side-shoots, called tillers, that grow along or just below the soil surface. Buds along the tiller produce new grass plants. The colony produced in this way binds with the soil to form turf which is hard wearing and is not destroyed by the feet of the grazing animals. These adaptations allow the grass to survive.

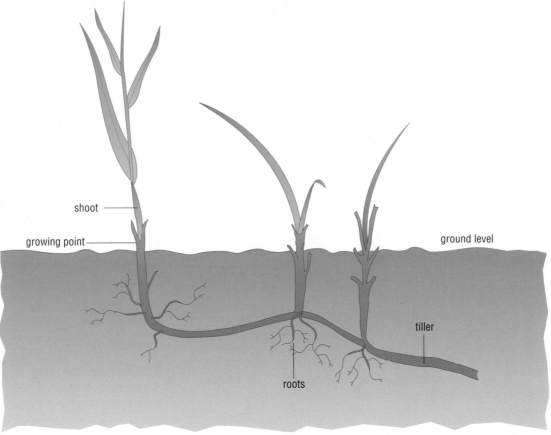

Figure 3.2 A grass plant.

Water plants

The roots of land plants have oxygen around them in the air spaces in the soil. In the waterlogged mud at the bottom of a pond there is very little oxygen for the root cells. The stems of water plants have cavities in them through which air can pass to the roots in order to overcome this problem.

Water plants use the gases they produce to hold their bodies up in the water and therefore do not need strong, supporting tissues like land plants. Minerals can be taken in from the water through the shoot surfaces of the water plant, leaving the root to act as an anchor. The leaves of submerged water plants are thin, allowing minerals in the water to pass into them easily. The leaves also have feathery structures that make a large surface area in contact with the water. This further helps the plant to take in all the essential minerals. Floating water plants like duckweed have a root that acts as a stabiliser.

3 In what ways are the features of a plant living in water different from a plant living on land?

Figure 3.3 A pond with a range of water plants.

Daily adaptations in plants

Flowers

Some plants such as the crocus and the tulip open their flowers during the day and close them at night. The flowers are open in the day so that insects may visit them for nectar, and in return transport pollen from one flower to another to bring about pollination. Many flowers close at night to protect the delicate structures inside the petals from low temperatures and from dew. The dew could wash the pollen off the stamens so that it cannot be picked up and transported by the insects.

Leaves

A few plants make movements of their leaves over a twenty-four hour period. Leaves not only make food but they also provide a large surface for the evaporation of water, which in turn helps to draw water through the plant from the roots. In clover, for example, each leaf is divided into three leaflets. During the day the leaflets spread out and become horizontal. In this position they are best placed for receiving sunlight to make food, and to lose water to the air which causes the plant to draw up water from the roots. The water is needed to make food, and it also helps to keep the leaves cool in the sunlight. In the evening the leaflets fold close together. This helps them to lose less water when the plant is not making food.

4 Do the petals of a crocus and the leaves of a clover plant move for the same reason? Explain your answer.

Seasonal adaptations in plants

The abiotic factors in a habitat change with the seasons. The grass plant is adapted to survive winter conditions but its short roots make it dependent upon the upper regions of the soil staying damp. In drought conditions the soil dries out and the grass dies. Daffodils are adapted to winter conditions as the leaves above the ground die and the plant forms a bulb in the soil.

Plants that float on the open water of a pond in spring and summer do not remain there in the winter. Duckweed produces individuals that sink to the pond floor. The plants around the water's edge die back and survive in the mud as thick stems called rhizomes.

5 In what ways are plants adapted to survive winter conditions?

Some adaptations in animals

A land animal

The tawny owl has several adaptations that allow it to catch mice at night. It has large eyes that are sensitive to the low intensity of light in the countryside at night. These allow it to see to fly safely. The edges of some of the owl's wing feathers are shaped to move noiselessly through the air when the bird beats its wings. This prevents the mouse's keen sense of hearing from detecting the owl approaching in flight. The owl has sharp talons on its toes that act as daggers, to kill its prey quickly and to help carry the prey away to be eaten at a safe perch.

Figure 3.4   A tawny owl, swooping down on a wood mouse.

6 What adaptations does the tawny owl have that allow it to detect its prey, approach its prey and attack its prey?

7 Why should the owl kill its prey quickly?

8 What adaptations do you think a mouse may have to help it survive a predator's attack?

A water animal

Although it lives underwater the diving beetle breathes air. It comes to the surface and pushes the tip of its abdomen out of the water. The beetle raises its wing covers and takes in air through breathing holes, called spiracles, on its back. (In insects living on land the spiracles are on the side of the body.) When the beetle lowers its wing covers more air is trapped in the hairs between them. It is able to breathe this air while it swims underwater.

Daily adaptation in animals

Animals are adapted to being active at certain times of the day and resting at other times. At night most birds roost (sleep) but as soon as it is light they may start flying about in search of food. They have large eyes and consequently have excellent vision, which is essential for them to fly, land and search for food.

Figure 3.5   A diving beetle feeding on an earthworm.

At night they are replaced in the air by bats. These animals roost in the day and come out at dusk to hunt for flying insects. Bats do not use their eyes but have developed an echo-location system. They send out very high-pitched squeaks that we cannot hear. These sounds reflect off all the surfaces around the bat and travel back to the bat's ears. The bat uses the information from these sounds to work out the distance, size and shape of objects around it. This allows the bat to fly safely and detect insects in the air, which it can swoop down on and eat.

Many insects such as butterflies, bees and wasps are active and fly during the day. At night moths take to the air to search for food.

While darkness is the major feature of a habitat at night there is also a second feature – humidity. At night all surfaces in the habitat cool down and the air cools down too. This causes water vapour in the air to condense and form dew. The increase in humidity is ideal for animals such as slugs and woodlice, which have difficulty retaining water in their bodies in dry conditions. They hide away in damp places during the day but as more places in the habitat become damp at night they become more active and roam freely searching for food. In the morning as the humidity decreases they hide away again somewhere damp.

Seasonal adaptations in animals

The roe deer (see Figure 3.6) lives in woodlands. In the spring and summer when the weather is warm it has a coat of short hair to keep it cool. In the autumn and winter it grows longer hair that traps an insulating layer of air next to its skin. This reduces the loss of heat from its body.

The stoat grows a white coat in the winter which loses less heat than its darker summer coat. The stoat preys on rabbits and its white coat may also give it some camouflage when the countryside is covered in snow.

The ptarmigan is about the size of a hen. It lives in the north of Scotland, northern Europe and Canada. In summer it has a brown plumage that helps it hide away from predators while it nests and rears its young. In winter it has a white plumage that reduces the heat lost from its body and gives it camouflage. Feathers grow over its toes and make its feet into snowshoes which allow it to walk across the snow without sinking.

> **9** Imagine that you are camping in a wood.
> **a)** What animals may you expect to see during the day?
> **b)** What animals are active in the evening?

> **10** How do the adaptations of
> **a)** the roe deer,
> **b)** the stoat and
> **c)** the ptarmigan help them survive in the winter?
> **11** How might their winter adaptations affect their lives if they kept them through the spring and summer?

Figure 3.6 A roe deer in summer (left) and a ptarmigan in winter (right).

SURVIVAL

Looking for links between animals and plants

All the individuals of a species in a particular habitat make up a group called a population. All the populations of the different species in the habitat make up a group called a community.

Living things in a community can be grouped according to how they feed. Plants make their own food by photosynthesis and by taking in minerals. They are called the producers of food. Animals that feed on plants are called primary consumers. They are also known as herbivores. Animals that feed on primary consumers are called secondary consumers. They are also known as carnivores. The highest level consumer in a food chain is called the top carnivore. Some animals such as bears feed on both plants and animals. They are called omnivores. An omnivore feeds as a primary consumer when it feeds on plants, and as a secondary or higher level consumer when it feeds on animals.

Figure 3.7 This diagram shows how the numbers of producers and consumers compare in a habitat. The bigger the block, the bigger the number of organisms.

12 In the food chain identify
   a) the primary consumer,
   b) the secondary consumer,
   c) the tertiary consumer.
13 In the food chain identify the herbivore and the carnivores.
14 Construct some food chains with humans in them.
15 In the food chains you have constructed are humans classed as herbivores, carnivores or omnivores?

Food chains

The information about how food passes from one species to another in a habitat is set out as a food chain. For example, it may be discovered that a plant is eaten by a beetle which in turn is eaten by a shrew and that the shrew is eaten by an owl. This information can be shown as follows:

plant → beetle → shrew → owl

Once a food chain has been worked out further studies can be done on the species in it.

Energy and the food chain

While animals need food substances from their meals to nourish their bodies they also need energy to keep their bodies alive. The source of energy for almost all food chains is the Sun (for more details on the energy from the Sun see Chapter 24). Some of the energy in sunlight is trapped in the leaves of plants and stored in food that the plant makes. When a herbivore eats the plant the energy is transferred to the body of the herbivore. Some of this energy is used up by the herbivore to keep it alive, but some is stored in its tissues. When a carnivore eats a herbivore it takes in the energy from the herbivore's body and uses it to keep itself alive. The food chain is an example of an energy transfer system.

Food webs

When several food chains are studied in the habitat some species may appear in more than one. For example, a badger eats blackberries and also eats snails. The two food chains it appears in are:

blackberry → badger

plant → snail → badger

The two food chains can then be linked together as:

blackberry → badger
⬈
plant → snail

When all the food chains in the habitat are linked up they make a food web. A food web shows the movement of food through a habitat. It can also be used to help predict what might happen if one of the links in a food web was absent. Look at Figure 3.8 and think of each animal shown there not just as one animal but as the whole population of that species of animal in the wood. If you think of each animal in the plural, such as voles and finches, it may help you think about animal populations.

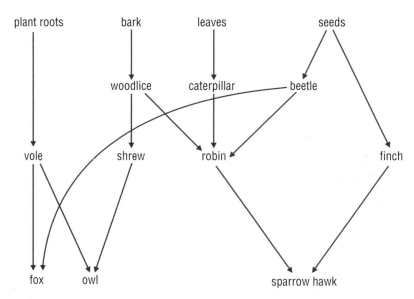

Figure 3.8   Examples of a food web in a woodland habitat.

Now imagine that the trees had a disease that made their leaves fall off. The caterpillars would starve and die, and they would not be available as food for the robins. This means that the robins must eat more beetles and woodlice if they are not to go hungry. The reduction in the number of woodlice would affect the shrews, as this is their only food. The beetle population would also fall, forcing the fox to search out more voles to eat.

16 Is a badger a herbivore, a carnivore or an omnivore? Explain your answer.

17 How may the numbers of other species in the wood change if each of the following was removed in turn:
a) fox,
b) seeds?

Summary

- There are two different kinds of factors in a habitat – abiotic and biotic factors (*see page 27*).
- Plants are adapted to their habitats (*see page 28*).
- Some plants are adapted to change during the day (*see page 29*).
- Plants are adapted to changes that occur with the seasons (*see page 29*).
- Animals are adapted to their habitats (*see page 30*).
- Animals are adapted to changes that occur during the day (*see pages 30–31*).
- Animals are adapted to changes that occur with the seasons (*see page 31*).
- Plants and animals are linked to each other by the way animals feed (*see page 32*).
- Food passes from one species to another along the food chain (*see page 32*).
- A food chain is an energy transfer system (*see page 32*).
- Food chains link together to form food webs (*see page 33*).

End of chapter questions

1 Construct a food web for the African plains from the following information.

Giraffes feed on trees, elephants feed on trees, eland feed on trees and bushes, hunting dogs feed on eland and zebra, finches feed on bushes, mice feed on roots, baboons feed on roots and locusts, gazelles, zebras and locusts feed on grass, foxes feed on mice, lions feed on eland, zebra and gazelle, hawks feed on finches, eagles feed on baboons, foxes and gazelles.

You may like to write the name of each living thing on a separate card and arrange the cards on a sheet of paper. You could write arrows on the paper between the cards but use pencil at first as you may find that you have to move the cards about to make the food web tidy.

2 Use the food web you have made for Question 1 to answer these questions.
   a) Which living organisms are producers?
   b) Which animals are herbivores?
   c) Which animals are carnivores?
   d) Which animal is an omnivore?

Looking at living things

You are probably aware that there are large numbers of different living things. Just think of the animals that you may see at a zoo, or the different plants that you may see in a park. You may also be aware that even among the same kind of living things there is variety. Just think of how you recognise different people by their own characteristic features. When scientists are looking at the features of living things they make careful observations and accurate records of what they see.

Making observations

When you make observations you look closely and with a purpose. For example, you may *look* at a plant and just see its flowers and leaves, but if you *observed* a plant you could study it to find out how the leaves and flowers are arranged on the stem. Leaves can be arranged in many ways, for example, they may grow alternately along a stem or they may be arranged in pairs. Flowers may be arranged singly or in columns.

Variation between species

Many living things have certain features in common. For example a cat, a monkey and a rabbit have ears and a tail. However, these features vary from one kind of animal to the next. In the species shown in the photographs in Figure 4.2, the external parts of the ears of the rabbit are longer than the ears of the cat. The external parts of the monkey's ears are on the side of its head while the other two animals have them on the top. The cat and the monkey have long tails but the monkey's tail is prehensile, which means it can wrap it around a branch for support while it hangs from a tree to collect fruit. (Only monkeys that come from South America have prehensile tails.) A rabbit's tail is much shorter than the cat's tail and the monkey's tail.

These variations in features are used to separate living things into groups and form a classification system which is used worldwide.

Variation within a species

The individuals in a species are not identical. Each one differs from all the others in many small ways. For example, one person may have dark hair, blue eyes and ears with lobes while another person may have fair

B   St John's wort

A   Rosebay willow herb

Figure 4.1   The leaves and flowers in two plants.

---

1 How are the leaves arranged in plant A and plant B in Figure 4.1?
2 How are the flowers arranged in plant A and plant B?

Figure 4.2   A cat, a rabbit and a South American monkey.

hair, brown eyes and ears without lobes. Another person may have different combinations of these features.

There are two kinds of variation that occur in a species. They are continuous variation and discontinuous variation.

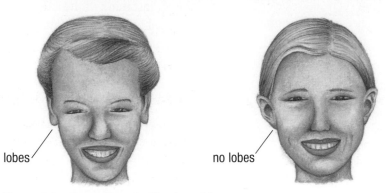

lobes   no lobes

Figure 4.3   Ears with and without ear lobes.

Continuous variation

A feature that shows continuous variation may vary in only a small amount from one individual to the next, but when the variations of a number of individuals are compared they form a wide range. Examples include the range of values seen in different heights or body masses.

Discontinuous variation

A feature that shows discontinuous variation shows a small number of distinct conditions, such as being male or female, or having ear lobes or no ear lobes. There is not a range of stages between the two as there is between a short person and a tall person. However, there are very few examples of discontinuous variation in humans.

The causes of variation

Some members of the family in Figure 4.4 have similar features. They are found in different generations which suggests that the features could be inherited. In fact we inherit many features, and the way they are passed on is explained in Chapter 9. Some variations may also be due to the environment.

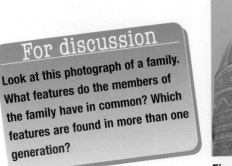

For discussion

Look at this photograph of a family. What features do the members of the family have in common? Which features are found in more than one generation?

Figure 4.4   Members of a family.

3 How else could the environment affect the development of an organism? Give another example for a plant and an animal.

Variation and the environment

The environment can affect the features of a living organism. For example, if some seedlings of a plant are grown in the dark and some in the light they will have different features. Those grown in the dark will be tall, spindly and have yellow leaves, while those grown in the light will have shorter, firmer stems with larger leaves that are green. Lack of food in the environment makes animals become thin. It can also slow down the growth of young animals.

Classifying living things

Living things are put into groups so that they can be studied more easily. The largest groups are called kingdoms. Scientists have now named five kingdoms. They are bacteria (Monera), single-celled organisms (Protoctista), fungi, plants and animals. Each kingdom contains a large number of living things that all have a few major features in common. Table 4.1 shows the features that are used to place living things in either the animal kingdom or the plant kingdom.

Table 4.1 The features of living things in the plant and animal kingdoms.

Plant kingdom	Animal kingdom
Make their own food from air, water, sunlight and chemicals in the soil	Cannot make their own food. Eat plants and animals
Body contains cellulose for support	Body does not contain cellulose
Have the green pigment chlorophyll	Do not have chlorophyll
Stay in one place	Move about

The way that the animal kingdom is divided up into subgroups is described on the following pages. A similar way of subgrouping is used to divide up the other kingdoms.

Subgroups of the animal kingdom

4 How is an earthworm different from a wasp, and how is it similar to a wasp?

Each kingdom is divided into groups called phyla (singular: phylum). Each phylum contains living things with more similarities. For example, in the phylum Arthropoda, which means 'jointed leg', all the animals have a skeleton on the outside of their body and have jointed legs. The phyla of the animal kingdom can be put into two groups called the invertebrates and the vertebrates. Invertebrates do not have an inside skeleton of cartilage or bone. Vertebrates do have an inside skeleton of cartilage or bone. The main phyla in the invertebrate group are the jellyfish, flatworms, annelid worms, arthropods, molluscs and echinoderms.

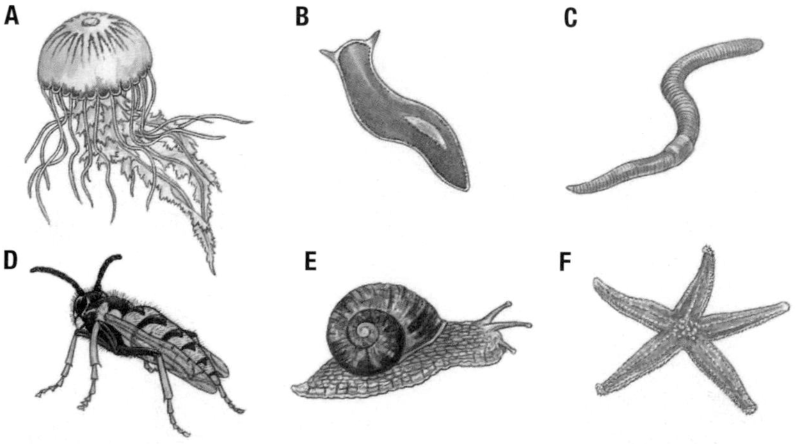

Figure 4.5   Examples of invertebrates. A: jellyfish (compass jellyfish), B: flatworm (pond flatworm), C: annelid worm (earthworm), D: arthropod (wasp), E: mollusc (snail), F: echinoderm (starfish).

There is only one phylum in the vertebrate group. It is known as the Chordata. The invertebrate and vertebrate groups are not actually part of the classification system but they are widely known and used to separate the animals in the Chordata from the animals in the other phyla.

Each phylum is divided up into groups called classes. The phylum Chordata contains seven classes that form the vertebrate group. The classes are jawless fish, cartilaginous fish, bony fish, amphibians, reptiles, birds, and mammals.

5 How is a goldfish different from a frog, and how is it similar?

Table 4.2   The features of five of the vertebrate classes.

Bony fish	Amphibians	Reptiles	Birds	Mammals
Scales, fins Eggs laid in water	Smooth skin Eggs laid in water	Scales. Soft-shelled eggs laid on land	Feathers Hard-shelled eggs	Hair Suckle young with milk

Setting up a system

Carl Linnaeus (1707–1778) worked out a way of describing how one kind of living thing was different from another. He then began putting very similar living things into the same group. He gave each living thing two names. The first name was the name of its genus and the second name was its own specific name or species name. The names were made from Latin and Greek. These were two languages that scientists of every country learnt, so everyone could understand them. The words usually described the appearance of a living thing. For example, the genus and species name of the African clawed toad is *Xenopus laevis*. Xenopus is made from two Greek words – xenos meaning 'strange' and pous meaning 'foot'. The words refer to the toad's webbed hind foot which has three toes, each capped with a dark, sharp claw. The word 'laevis' is Latin for smooth and refers to the toad's smooth skin.

For discussion

Why were the common names or local names not used in the scientific naming of plants and animals?

Other kingdoms of living things

As well as the animals, there are four other kingdoms of living things. They are the plants, Monera, Protoctista and fungi. You can find out about the plant kingdom in Chapter 11 and the other kingdoms in Chapter 7.

Keys

The way in which living things are divided up into groups can be used to identify them. The features of organisms can be set out as a spider key or organised as pairs of statements in a numbered key.

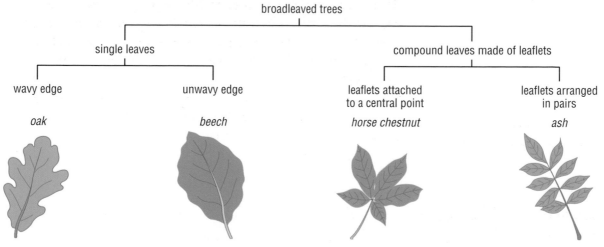

broadleaved trees

single leaves — compound leaves made of leaflets

wavy edge — unwavy edge — leaflets attached to a central point — leaflets arranged in pairs

oak — *beech* — *horse chestnut* — *ash*

Figure 4.6 Spider key of leaves from broadleaved trees.

6 Make a spider key for the four animals in Figure 4.7. Look carefully at the animals. Choose a feature they have all got in common to start at the top of the key.

A

B

C

D

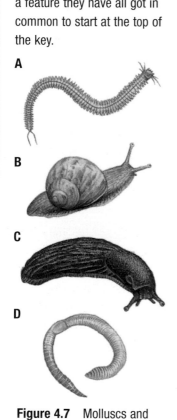

Figure 4.7 Molluscs and annelid worms.

Spider key

On each 'leg' of the spider is a feature that is possessed by the living thing below it. An example is shown in Figure 4.6. A spider key is read by starting at the top in the centre and reading the features down the legs until the specimen is identified.

Numbered key

You work through a numbered key by reading each pair of statements and matching the description of one of them to the features you see on the specimen you are trying to identify. At the end of each statement there is an instruction to move to another pair of statements or to the name of a living thing. Here is a simple numbered key. It can be used to identify molluscs that live in freshwater habitats such as rivers, lakes and ponds.

1 a) Single shell .... see 2
 b) Two shells .... see 6
2 a) Snail with a plate that closes the shell mouth .... *Bithynia*
 b) Snail without a plate that closes the shell mouth .... see 3
3 a) Snail without a twisted shell .... Freshwater .... limpet
 b) Snail with a twisted shell........................... see 4
4 a) Shell in a coil .... Ramshorn .... snail
 b) Shell without a coil.................................... see 5
5 a) Snail with triangular tentacles.................... Pond snail
 b) Snail with long, thin tentacles .... Bladder snail
6 a) Animal has threads attaching it to a surface.. Zebra mussel
 b) Animal does not have threads attaching it to a surface .... see 7
7 a) Shell larger than 25 mm .... Freshwater .... mussel
 b) Shell smaller than 25 mm........................... Pea mussel

7 Identify the molluscs in Figure 4.8 using the numbered key on page 39. In each case write down the number of each statement you followed to make the identification. For example, specimen **A** is identified by following statements **1a**, **2b**, **3a**. It is a freshwater limpet.

8 Why should another feature in addition to size be added to the statements in part 7 of the key?

Figure 4.8 Some freshwater molluscs.

9 Make up a numbered key to identify the arthropods in Figure 4.9. Begin by
separating the butterfly, which has six legs, from the others.

A

millipede

B

centipede

C

crab

D

spider

E

butterfly

Figure 4.9 Arthropod specimens.

Summary

- Detailed features of living things are seen by close observation and accurate recording (*see page 35*).
- There is variation between species (*see page 35*).
- There is variation within a species (*see page 35*).
- The environment can affect variation in a species (*see page 37*).
- Living things are classified by putting them into groups. The groups include kingdom, phylum, class, genus and species (*see pages 37–38*).
- Keys are used to identify living things (*see pages 38–39*).

End of chapter questions

1 What kind of living organisms are the following:
 a) does not have a backbone but has five arms,
 b) has a backbone and wings,
 c) does not have a backbone but has wings,
 d) has scales and lays eggs in water,
 e) has scales and lays eggs on land,
 f) has a backbone and hair?

2 Imagine that you have landed on a distant planet. When you climb out of your spacecraft you find some small eight-legged, six-eyed animals leaping about. You call them hoppys, and gather the following information about twenty of them.
 a) Arrange the hoppys into five size groups based on their mass.
 b) Display the numbers in the groups in a bar chart.
 c) Arrange the hoppys into groups based on colour.
 d) Display the numbers in the groups in a bar chart.
 e) Which feature shows continuous variation?
 f) Which feature shows discontinuous variation?
 g) Is there any relationship between the mass of the hoppys and their colours? Describe what you find.
 h) Speculate on a reason for your findings.

Table 4.3

Hoppy	Mass (g)	Colour
1	200	green
2	349	green
3	210	green
4	615	blue
5	430	yellow
6	570	red
7	402	yellow
8	429	yellow
9	317	green
10	520	red
11	460	yellow
12	403	yellow
13	330	green
14	489	yellow
15	502	red
16	630	blue
17	410	yellow
18	380	green
19	550	red
20	445	yellow

Food and digestion

Figure 5.1 Food for sale at a school canteen.

1 Write a description of your daily eating pattern.
2 Compare your pattern with the two on this page. Which one does your pattern resemble?
3 From what you already know, try to explain which diet is healthier.

For discussion

How healthy is your eating pattern? What changes would make it healthier? Do other people agree?

Some children do not eat breakfast. They have some sweets or crisps on the way to school. At break they eat a chocolate bar or have a fizzy drink. At lunch time they always have chips with their meal. In the afternoon they have some more sweets and for their evening meal they avoid green vegetables. Through the evening they have snacks of sweets, crisps and fizzy drinks.

Other children eat a breakfast of cereals and milk, toast and fruit juice. They eat an apple at break and have a range of lunch time meals through the week which include different vegetables, pasta and rice. In the afternoon they may have an orange and eat an evening meal with green vegetables. They may have a milky drink at bed time.

Nutrients

A chemical that is needed by the body to keep it in good health is called a nutrient. The human body needs a large number of different nutrients to keep it healthy. They can be divided up into the following nutrient groups:

- carbohydrates
- proteins
- minerals
- fats
- vitamins

In addition to these nutrients the body also needs water. It accounts for 70% of the body's weight and provides support for the cells, it carries dissolved materials around the body and it helps in controlling body temperature.

Carbohydrates

Carbohydrates are made from the elements carbon, hydrogen and oxygen. The atoms of these elements are linked together to form molecules of sugar. There are different types of sugar molecule but the most commonly occurring is glucose. Glucose molecules link together in long chains to make larger molecules, such as starch. Glucose and starch are two of the most widely known carbohydrates but there are others, such as cellulose, which is an important component of dietary fibre.

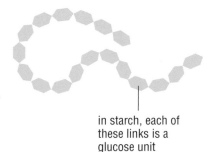

in starch, each of these links is a glucose unit

Figure 5.2 Carbohydrate molecule.

43

Fats

Fats are made of large numbers of carbon and hydrogen atoms linked together into long chains together with a few oxygen atoms. There are two kinds of fats – the solid fats produced by animals, such as lard, and the liquid fat or oil produced by plants, such as sunflower oil.

Proteins

Proteins are made from atoms of carbon, hydrogen, oxygen and nitrogen. Some proteins also contain sulphur and phosphorus. The atoms of these elements join together to make molecules of amino acids. Amino acids link together into long chains to form protein molecules.

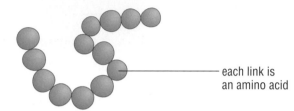

each link is an amino acid

Figure 5.3   Protein molecule.

Vitamins

Unlike carbohydrates, fats and proteins, which are needed by the body in large amounts, vitamins are only needed in small amounts. When the vitamins were first discovered they were named after letters of the alphabet. Later, when the chemical structure of their molecules had been worked out, they were given chemical names.

Minerals

The body needs 20 different minerals to keep healthy. Some minerals, such as calcium, are needed in large amounts but others, such as zinc, are needed in only tiny amounts and are known as trace elements.

How the body uses nutrients

Carbohydrates

Carbohydrates contain a large amount of energy that can be released quickly inside the body. They are used as fuels to provide the energy for keeping the body alive. Cellulose, which makes up the walls of plant cells, is a carbohydrate. We cannot digest it but its presence in our food gives the food a solid property. This allows the muscles of the gut to push the food along, aiding digestion and preventing constipation. Cellulose in food is known as dietary fibre.

Fats

Fats are needed for the formation of cell membranes. They also contain even larger amounts of energy than carbohydrates. The body cannot release the energy in fats as quickly as the energy in carbohydrates, so fats are used to store energy. In mammals the fat forms a layer under the

4 What elements are found in carbohydrates, fats and proteins?

5 Which two words are used to describe the structure of carbohydrate, fat and protein molecules?

6 A science teacher held up a necklace of beads to her class and said it was a model of a protein molecule. What did each bead represent?

NUTRITION INFORMATION		Typical value per 100g	Per 30g Serving with 125ml of Semi-Skimmed Milk
ENERGY	kJ	1550	700*
	kcal	370	170
PROTEIN	g	8	7
CARBOHYDRATE	g	83	31
(of which sugars)	g	(8)	(9)
(starch)	g	(75)	(22)
FAT	g	0.7	2.5*
(of which saturates)	g	(0.2)	(1.5)
FIBRE	g	3	0.9
SODIUM	g	1.1	0.4
VITAMINS:		(%RDA)	(%RDA)
THIAMIN (B₁)	mg	1.2 (85)	0.4 (30)
RIBOFLAVIN (B₂)	mg	1.3 (85)	0.6 (40)
NIACIN	mg	15 (85)	4.6 (25)
VITAMIN B₆	mg	1.7 (85)	0.6 (30)
FOLIC ACID	μg	333 (165)	110 (55)
VITAMIN B₁₂	μg	0.85 (85)	0.75 (75)
IRON	mg	7.9 (55)	2.4 (17)

Figure 5.4   The nutrients in a food product are displayed on the side of the packet.

skin. This acts as a heat insulator and helps to keep the mammal warm in cool conditions. Many mammals increase their body fat in the autumn so that they can draw on the stored energy if little food can be found in the winter. Some plants store oil in their seeds.

Proteins

Proteins are needed for building the structures in cells and in the formation of tissues and organs. They are needed for the growth of the body, to repair damaged parts, such as cut skin, and to replace tissues that are constantly being worn away, such as the lining of the mouth.

Chemicals that take part in the reactions for digesting food and in speeding up reactions inside cells are called enzymes. These are also made from proteins.

Vitamins

Each vitamin has one or more uses in the body.

Table 5.1   Vitamins and their uses.

Vitamin	Effect on body	Good sources
A	Increased resistance to disease Helps eyes to see in the dark	Milk, liver, cod-liver oil
B1	Prevents digestive disorders Prevents the disease beriberi	Bread, milk
C	Prevents the disease scurvy in which gums bleed and the circulatory system is damaged	Blackcurrant, orange, lemon
D	Prevents the disease rickets in which bones become soft and leg bones of children may bend	Egg yolk, butter, cod-liver oil

A lack of vitamin C causes the deficiency disease called scurvy. As the disease develops, bleeding occurs at the gums in the mouth, under the skin and into the joints. Death may occur due to massive bleeding in the body.

7 A meal contains carbohydrate, fat, protein, vitamin C, calcium and iron. What is the fate of each of these substances in the body?

8 Which carbohydrate cannot be digested by humans and how does it help the digestive system?

9 A seal is a mammal. How can it survive in the cold polar seas when a human would die in a few minutes?

Figure 5.5   Seals on the ice.

Minerals

Each mineral may have more than one use. For example, calcium is needed to make strong bones and teeth. It is also used to make muscles work and for blood to clot. A lack of calcium in the diet can lead to weak bones and high blood pressure. The mineral iron is used to make the red blood pigment called haemoglobin.

Water

About 70% of the human body is water. The body can survive for only a few days without a drink of water.

Every chemical reaction in the body takes place in water. The blood plasma is made mainly from water. It is the liquid that transports all the other blood components around the body.

Water is used to cool down the body by the evaporation of sweat from the skin.

Table 5.2 The nutrients in some common foods.

Food (100 g)	Protein (g)	Fat (g)	Carbohydrate (g)	Calcium (mg)	Iron (mg)	Vitamin C (mg)	Vitamin D (µg)
Potato	2.1	0	18.0	8	0.7	8–30	0
Carrot	0.7	0	5.4	48	0.6	6	0
Bread	9.6	3.1	46.7	28	3.0	0	0
Spaghetti	9.9	1.0	84.0	23	1.2	0	0
Rice	6.2	1.0	86.8	4	0.4	0	0
Lentil	23.8	0	53.2	39	7.6	0	0
Pea	5.8	0	10.6	15	1.9	25	0
Jam	0.5	0	69.2	18	1.2	10	0
Peanut	28.1	49.0	8.6	61	2.0	0	0
Lamb	15.9	30.2	0	7	1.3	0	0
Milk	3.3	3.8	4.8	120	0.1	1	0.05
Cheese 1	25.4	35.4	0	810	0.6	0	0.35
Cheese 2	15.3	4.0	4.5	80	0.4	0	0.02
Butter	0.5	81.0	0	15	0.2	0	1.25
Chicken	20.8	6.7	0	11	1.5	0	0
Egg	12.3	10.9	0	54	2.1	0	1.50
Fish 1	17.4	0.7	0	16	0.3	0	0
Fish 2	16.8	18.5	0	33	0.8	0	22.20
Apple	0.3	0	12.0	4	0.3	5	0
Banana	1.1	0	19.2	7	0.4	10	0
Orange	0.8	0	8.5	41	0.3	50	0

Notes for Tables 5.2 and 5.3

Vegetables are raw; the bread is wholemeal bread; cheese 1 is cheddar cheese; cheese 2 is cottage cheese; fish 1 is a white fish, such as cod; fish 2 is an oily fish, such as herring.

10 In Table 5.2, which foods contain the most
   a) protein,
   b) fat,
   c) carbohydrate,
   d) calcium,
   e) iron,
   f) vitamin C,
   g) vitamin D?
11 Which foods would a vegetarian not eat?
12 Which foods would a vegetarian have to eat more of and why?
13 Which food provides all the nutrients?
14 Why might you expect this food to contain so many nutrients?

The amounts of nutrients in food

The amounts of nutrients in foods have been worked out by experiment and calculation. The amounts are usually expressed for a sample of food weighing 100 g. Table 5.2 shows the nutrients in a small range of common foods.

Keeping a balance

In order to remain healthy the diet has to be balanced with the body's needs. A balanced diet is one in which all the nutrients are present in the correct amounts to keep the body healthy. You do not need to know the exact amounts of nutrients in each food to work out whether you have a healthy diet. A simple way is to look at a chart showing food divided into groups, with the main nutrients of each group displayed (see Table 5.4, overleaf). You can then see if you eat at least one portion from each group each day and more portions of the food groups that lack fat. Remember that you also need to include fibre even though it is not digested. It is essential for the efficient working of the muscles in the alimentary canal. Fibre is found in cereals, vegetables and pulses, such as peas and beans.

15 Table 5.3 shows the amount of energy provided by 100 g of each of the foods shown in Table 5.2. Arrange the nine highest energy foods in order starting with the highest and ending with the lowest. Look at the nutrient contents of these foods in Table 5.2.
   a) Do you think the energy is stored as fat or as carbohydrate in each of the nine highest energy foods?
   b) Arrange the foods into groups according to where you think the energy is stored.
   c) Do the foods you have identified store the same amount of energy? (See also page 44.) Explain your answer.
16 Why might people who are trying to lose weight eat cottage cheese instead of cheddar cheese?
17 Mackerel is an oily fish. Describe the nutrients you would expect it to contain.

Table 5.3 The energy value of some common foods.

Food (100 g)	Energy (kJ)
Potato	324
Carrot	98
Bread	1025
Spaghetti	1549
Rice	1531
Lentil	1256
Pea	273
Jam	1116
Peanut	2428
Lamb	1388
Milk	274
Cheese 1	1708
Cheese 2	480
Butter	3006
Chicken	602
Egg	612
Fish 1	321
Fish 2	970
Apple	197
Banana	326
Orange	150

18 Look again at the eating pattern you prepared for Question 1 on page 43. Analyse your diet into the food groups shown in Table 5.4. How well does your diet provide you with all the nutrients you need?

19 Table 5.5 shows how the average energy requirements of males and females change from the age of 2 to 25 years. Plot graphs of the information given in the table.

20 Describe what the graphs show.

21 Explain why there is a difference in energy needs between a 2-year-old child and an 8-year-old child.

22 Explain why there is a difference in energy needs between an 18-year-old male and an 18-year-old female.

23 Explain why there is a change in the energy needs as a person ages from 18 to 25.

24 What changes would you expect in the energy used by:
   a) a 25-year-old person who changed from a job delivering mail to working with a computer,
   b) a 25-year-old person who gave up working with computers and took a job on a building site that involved carrying heavy loads,
   c) a 25-year-old female during pregnancy?

25 What happens in the body if too much fat, carbohydrate or protein is eaten?

26 Why do people become thin if they do not eat enough high-energy food?

Table 5.4 The groups of foods and their nutrients.

Vegetables and fruit	Cereals	Pulses	Meat and eggs	Milk products
Carbohydrate	Carbohydrate	Carbohydrate	Protein	Protein
Vitamin A	Protein	Protein	Fat	Fat
Vitamin C	B vitamins	B vitamins	B vitamins	Vitamin A
Minerals	Minerals	Iron	Iron	B vitamins
Fibre	Fibre	Fibre		Vitamin C
				Calcium

Table 5.5 Average daily energy used by males and females.

| Age (years) | Daily energy used (kJ) | |
	Male	Female
2	5500	5500
5	7000	7000
8	8800	8800
11	10 000	9200
14	12 500	10 500
18	14 200	9600
25	12 100	8800

Malnutrition

If the diet provides too few nutrients or too many nutrients malnutrition occurs. Lack of a nutrient in a diet may produce a deficiency disease, such as scurvy or anaemia. Scurvy is a deficiency disease caused by a lack of vitamin C and anaemia is a deficiency disease caused by a lack of iron.

Too much high-energy food such as carbohydrate and fat leads to the body becoming overweight. If the body is extremely overweight it is described as obese. If too little high-energy food is eaten the body becomes thin because it uses up energy stored as fat. Energy stored in protein in the muscles can also be used up.

The condition anorexia nervosa can lead to extreme weight loss and possibly death. It occurs mainly in teenage girls but is occurring increasingly in teenage boys and adult men and women. People suffering from anorexia nervosa eat very little and fear gaining weight. As soon as the condition is diagnosed, they need careful counselling to stand the best chance of making a full recovery.

A healthy diet

The body needs a range of nutrients to keep healthy and everyone should eat a balanced diet to provide these nutrients. Regular eating of high-energy snacks, such as sweets, chocolate, crisps, ice-cream and chips, between meals unbalances the diet and can lead to the body becoming overweight, damage to the teeth (see page 68) and ill-health. Overweight people have to make more effort than normal to move so they tend to take less exercise. In time this can affect the heart (see page 96).

There are alternatives to high-energy snacks. These are fruits and raw vegetables, such as celery, tomatoes and carrots. In addition to being lower in energy they also provide more vitamins and minerals.

Digestion

Your food comes from the tissues of animals and plants. To enter the cells of your body these tissues have to be broken down. This releases the nutrients (carbohydrates, proteins, fats, minerals and vitamins). Some of them are in the form of big molecules. They must be broken down into smaller molecules that dissolve in water and can pass through the wall of the gut. This process of making the food into a form that can be taken into the body is called digestion. It takes place in the digestive system, which is made up of the alimentary canal and organs such as the liver and pancreas.

The breakdown of food

There are two major processes in the breakdown of food: physical and chemical. Food is physically broken down from large objects into small objects in the mouth. Chemical breakdown begins in the mouth and continues along the alimentary canal.

The physical breakdown of food

The teeth play a major part in the physical breakdown of food. There are four kinds of teeth. The chisel-shaped incisor teeth are at the front of the mouth. These are for biting into soft foods like fruits. Next to the incisors are the canines. These are pointed, and in dogs and cats they form the fang teeth that are used for tearing into tougher food like meat. Humans do not eat much tough food so they use their canines as extra incisors. The premolars and the molars have raised parts called cusps

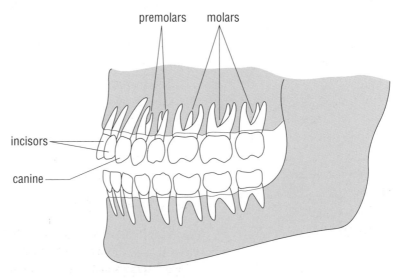

Figure 5.6 The four types of teeth in the mouth.

with grooves between them. They form a crushing and grinding surface at the back of the mouth. The action of the teeth breaks up the food into small pieces.

The chemical breakdown of food

Proteins, fats and carbohydrates are large molecules which are made from smaller molecules linked together. The large molecules do not dissolve in water and cannot pass through the lining of the digestive system into the body. The smaller molecules from which they are made, however, *do* dissolve in water and *do* pass through the wall of the digestive system. Almost all reactions in living things involve chemicals called enzymes. They are made by the body from proteins and they speed up chemical reactions. Digestive enzymes speed up the breakdown of the large molecules into smaller ones.

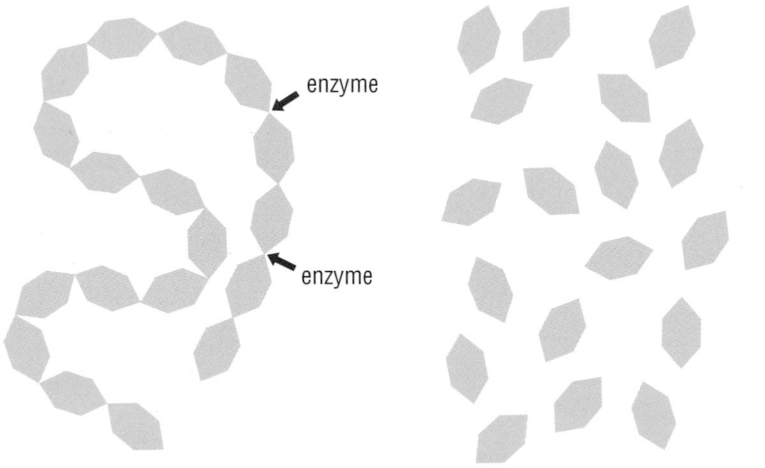

27 Which smaller molecules join together to form
   a) carbohydrates,
   b) proteins? (See also page 44.)
28 What do enzymes in the digestive system do?

a large molecule is attacked by an enzyme

it is broken into smaller molecules

Figure 5.7   The action of an enzyme on a large food molecule.

Along the alimentary canal

When your mouth waters

The liquid that occurs in your mouth is called saliva. You can make up to $1\frac{1}{2}$ litres of saliva in 24 hours. Saliva is made by salivary glands. The glands are made up of groups of cells that produce the saliva, and ducts (tubes) that deliver it to the mouth.

Saliva is 99% water but it also contains a slimy substance called mucin and an enzyme called amylase which begins the digestion of starch in the food. The mucin coats the food and makes it easier to swallow. Amylase begins the breakdown of starch molecules into sugar molecules.

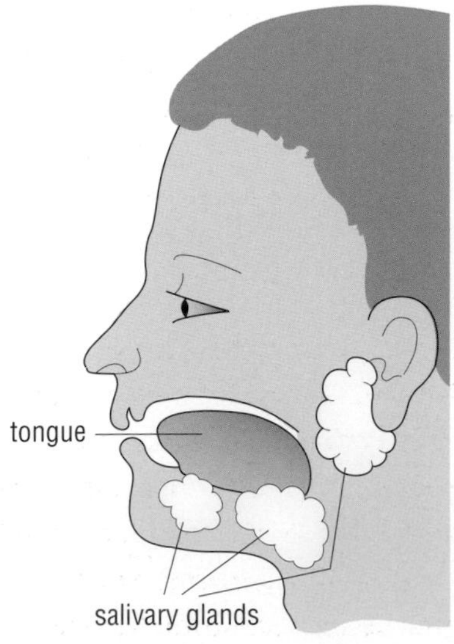

When you swallow

When you have chewed your food, it is made into a pellet which is pushed to the back of your mouth by your tongue. Swallowing causes the food to slide down your oesophagus (gullet), which is the tube connecting the mouth to the stomach. It has two layers of muscles in its walls: longitudinal and circular (see Figure 5.9).

tongue

salivary glands

Figure 5.8   The location of the salivary glands.

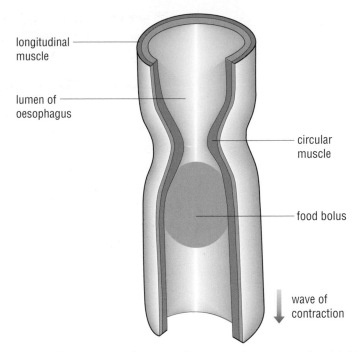

Figure 5.9   The structure of the oesophagus and the process of peristalsis.

In the oesophagus, when the circular muscles contract, they squeeze on the food and push it along the tube. The longitudinal muscles then contract to stretch the circular muscles once again. The circular muscles do not all contract at the same time. Those at the top of the oesophagus contract first then a region lower down follows and so on until the food is pushed into the stomach. This wave of muscular contraction is called peristalsis. Peristaltic waves also occur in other parts of the alimentary canal to push the food along.

Stomach

The stomach wall is lined with glands. These produce hydrochloric acid and a protein-digesting enzyme called pepsin. The hydrochloric acid kills many kinds of bacteria in the food and provides the acid conditions that pepsin needs to start breaking down protein in the food.

The food is churned up by the action of the muscles. It is prevented from leaving the stomach by a valve. When it has been broken down into a creamy liquid, the valve opens which allows the liquid food to pass through into the next part of the digestive system.

Duodenum, liver and pancreas

The duodenum is a tube that connects the stomach to the small intestine (see Figure 5.11, overleaf). Two other tubes are connected to it. One tube carries a green liquid called bile from the gall bladder to mix with the food. Bile is made in the liver. It helps break down fat into small droplets so that fat-digesting enzymes can work more easily. The second tube comes from the pancreas. This is a gland that produces a juice containing enzymes that digest proteins, fats and carbohydrates. The mixture of liquids from the stomach, liver and pancreas pass on into the small intestine.

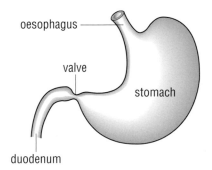

Figure 5.10   The stomach.

29 How does saliva help in digesting food?
30 What is peristalsis?
31 What does hydrochloric acid do?

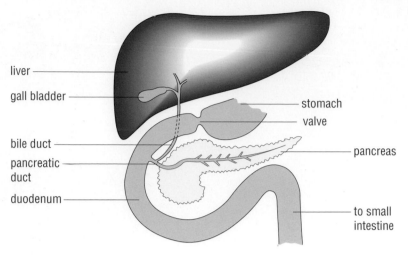

liver
gall bladder
stomach
valve
bile duct
pancreatic duct
duodenum
pancreas
to small intestine

Figure 5.11   The duodenum, liver and pancreas.

> **32** Where is bile made and what does it do?

Small intestine

In the small intestine enzymes complete the digestion of food. Proteins are broken down into amino acids, carbohydrates are broken down into sugars, and fats are broken down into fatty acids and glycerol. All of these products are small, soluble molecules that can pass through the wall of the small intestine. They are carried by the blood to all cells of the body.

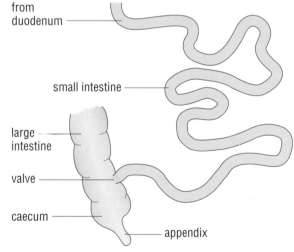

from duodenum
small intestine
large intestine
valve
caecum
appendix

Figure 5.12   The small intestine.

> **33** What are proteins, fats and carbohydrates broken down into?
> **34** Where are the digested foods absorbed?

Fate of undigested food

Indigestible parts of the food, such as cellulose, pass on through the small intestine to the large intestine (Figure 5.13). Here water and some dissolved vitamins are absorbed and taken into the body. The remaining semi-solid substances form the faeces which are stored in the rectum. The faeces are removed from the body through the anus perhaps once or twice a day in a process called egestion.

> **35** What happens to undigested food in the large intestine?
> **36** What happens in egestion?

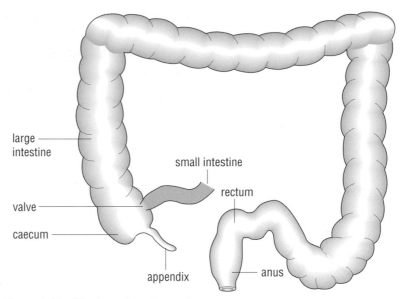

Figure 5.13 The large intestine and rectum.

Enzymes

An enzyme that digests carbohydrate is called a carbohydrase. An enzyme that digests protein is called a protease. An enzyme that digests fat is called a lipase.

Table 5.6 Enzymes.

Region of production	Kind of enzyme	Notes
Salivary glands in mouth	Carbohydrase	Enzyme is called salivary amylase
Gastric glands in stomach	Protease	Enzyme is called pepsin Hydrochloric acid is also made to help the enzyme work
Pancreas	Protease, carbohydrase, lipase	Enzymes enter the duodenum and mix with food and bile

37 What kind of enzyme is produced in
  a) the mouth,
  b) the stomach?
38 What kind of enzyme does bile help?
39 Where does bile come from?
40 Which organ of the digestive system produces all three kinds of enzyme?
41 Why do small droplets of fat get broken down by enzymes more quickly than large droplets?

Summary

- A chemical that is needed by the body to keep it healthy is called a nutrient (*see page 43*).
- The groups of nutrients are carbohydrates, fats, proteins, vitamins and minerals (*see pages 43–44*).
- Each nutrient has a specific use in the body (*see pages 44–46*).
- Different foods have different amounts of nutrients (*see page 47*).
- A balanced diet needs to be eaten for good health (*see page 47*).
- Water and fibre are essential components of the diet (*see pages 46 and 47*).
- The purpose of digestion is to break down the food into substances that can be absorbed and used by the body (*see page 49*).
- There are four kinds of teeth. They are the incisors, canines, premolars and molars. They have special shapes for specific tasks (*see page 49*).

(continued)

- Enzymes break down the large molecules in food into smaller molecules so that they can be absorbed by the body (*see page 50*).
- The food is moved along the gut by a wave of muscular contraction called peristalsis (*see page 51*).
- The food is digested by enzymes that are made in the salivary glands, the stomach wall, the pancreas and the wall of the small intestine (*see pages 50–53*).
- The liver produces bile which helps in the digestion of fat (*see page 51*).
- Digested food is absorbed in the small intestine (*see page 52*).
- The undigested food has water removed from it in the large intestine and is then stored in the rectum before being released through the anus (*see page 52*).

End of chapter questions

1 What is a healthy diet?

2 Collect a copy of Figure 5.14 from your teacher and use the other diagrams in this chapter to label all the parts of the digestive system.

3 Describe the digestion of a chicken sandwich.

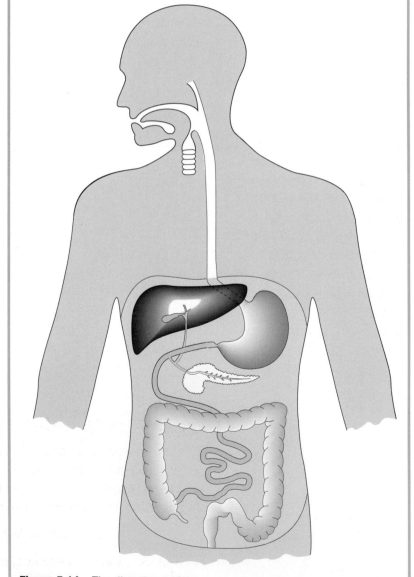

Figure 5.14 The digestive system.

Respiration

All life processes require energy. The energy is stored in food molecules and is released in respiration. In the human body, a sugar called glucose is the main source of energy. Most of it is formed by the digestion of starch. It dissolves in the blood and is transported to the cells.

Detecting the release of energy

When energy changes from stored energy in food to movement energy, some energy is released as heat. This is why you become hot when you take part in sport.

The production of heat can be used to indicate that something is alive. Seeds store energy and materials for plant growth. When seeds take in water the stored energy is used to power the growth of the plant inside, and germination occurs.

The production of heat by germinating seeds can be demonstrated in the following way. Two flasks are set up as shown in Figure 6.1.

<table>
<tr><td>1 Would you expect a rise in temperature if you put dead mouldy peas in a flask? Explain your answer.</td></tr>
<tr><td>2 In the burning process, a fuel takes part in a chemical reaction with oxygen and produces the same products as in respiration. How is burning different from respiration?</td></tr>
<tr><td>3 What would happen if burning occurred in the body instead of respiration?</td></tr>
</table>

thermometer

vacuum flask

living pea seeds

boiled pea seeds

Figure 6.1 Investigating heat produced by germinating seeds.

After 24 hours the temperature inside each flask is measured. The flask containing the living pea seeds will be found to be at a higher temperature than the other flask. This indicates that the living pea seeds are respiring.

Aerobic respiration

Glucose takes part in a chemical reaction with oxygen inside the cell. During this reaction glucose is broken down to carbon dioxide and water, and energy is released. This process is called aerobic respiration. The reaction can be written as a word equation:

glucose + oxygen → carbon dioxide + water + energy

The energy is released slowly in a series of stages during respiration.

Respiratory system

In Chapter 5 we saw how food is broken down in the digestive system in readiness for transport to the cells. In this section we look at how the respiratory system provides a means of exchanging the respiratory gases – oxygen and carbon dioxide.

The function of the respiratory system is to provide a means of exchanging oxygen and carbon dioxide that meets the needs of the body, whether it is active or at rest. In humans the system is located in the head, neck and chest. It can be divided into three parts – the air passages and tubes, the pump that moves the air in and out of the system, and the respiratory surface. The terms respiration and breathing are often confused but they do have different meanings. Breathing just describes the movement of air in and out of the lungs. Respiration covers the whole process by which oxygen is taken into the body, transported to the cells and used in a reaction with glucose to release energy, with the production of water and carbon dioxide as waste products.

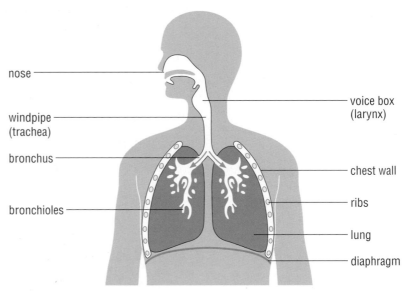

nose

voice box (larynx)

windpipe (trachea)

bronchus

chest wall

ribs

bronchioles

lung

diaphragm

Figure 6.2   Respiratory system.

Air passages and tubes

Nose

Air normally enters the air passages through the nose. Hairs in the nose trap some of the dust particles that are carried on the air currents. The lining of the nose produces a watery liquid called mucus. This makes the air moist as it passes inwards and also traps bacteria that are carried on the air currents. Blood vessels beneath the nasal lining release heat that warms the air before it passes into the lungs.

Windpipe

The windpipe or trachea is supported by rings of cartilage. Each ring is in the shape of a 'C'. The inner lining of the windpipe has two types of cells. They are mucus-secreting cells and ciliated epithelial cells. Dust particles and bacteria are trapped in the mucus. The cilia beat backwards and forwards to move the mucus to the top of the windpipe where it enters the back of the mouth and is swallowed.

Bronchi and bronchioles

The windpipe divides into two smaller tubes called bronchi. (This is the name for more than one tube. A single tube is called a bronchus.) The two bronchi are also supported by rings of cartilage and have the same lining as the windpipe.

The bronchi divide up into many smaller tubes called bronchioles. These have a diameter of about 1 mm. The bronchioles divide many times. They have walls made of muscle but no cartilage. The wall muscles can make the bronchiole diameter narrower or wider.

Air pump

The two parts of the air pump are the chest wall and the diaphragm. They surround the cavity in the chest. Most of the space inside the chest is taken up by the lungs. The outer surfaces of the lungs always lie close to the inside wall of the chest. The small space between the lungs and the chest wall is called the pleural cavity. The cavity contains a film of liquid that acts like a lubricating oil, helping the lung and chest wall surfaces to slide over each other during breathing.

Chest wall

This is made by the ribs and their muscles. Each rib is attached to the backbone by a joint that allows only a small amount of movement. The muscles between the ribs move the rib cage up and down.

Diaphragm

This is a large sheet of muscle which separates the chest cavity, containing the lungs and heart, from the lower body cavity, containing the stomach, intestines, liver, kidneys and reproductive organs. When the diaphragm muscle contracts, it pulls the diaphragm down.

Respiratory surface

At the end of each bronchiole are bubble-like structures called alveoli. Each alveolus has a moist lining, a thin wall and is supplied with tiny blood vessels called capillaries.

Oxygen from the inhaled air dissolves in the moist alveolar lining and moves by diffusion through the walls of the alveolus and the capillary

> **4** What structures hold the air passages open in the windpipe and bronchi?

> **5** Why is there a film of liquid in the pleural cavity?

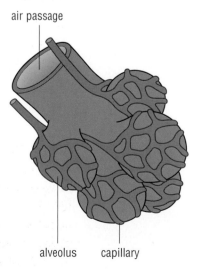

air passage

alveolus   capillary

Figure 6.3 Alveoli.

next to it. The oxygen diffuses into the blood and enters the red blood cells, which contain a dark red substance called haemoglobin. The oxygen then combines with the haemoglobin to make oxyhaemoglobin, which is bright red. Blood that has received oxygen from the air in the lungs is known as oxygenated blood.

Carbon dioxide is dissolved in the watery part of the blood called the plasma. It moves by diffusion through the capillary and alveolar walls and changes into a gas as it leaves the moist lining of the alveolus.

Blood moves through the capillaries very quickly, so a large amount of oxygen and carbon dioxide can be exchanged in a short time.

The spongy structure of the lungs is produced by the 300 million alveoli which make a very large surface area through which the gases can be exchanged. It is like having the surface area of a tennis court wrapped up inside two footballs! If this surface area is reduced then health suffers (see Smoking and health, page 93).

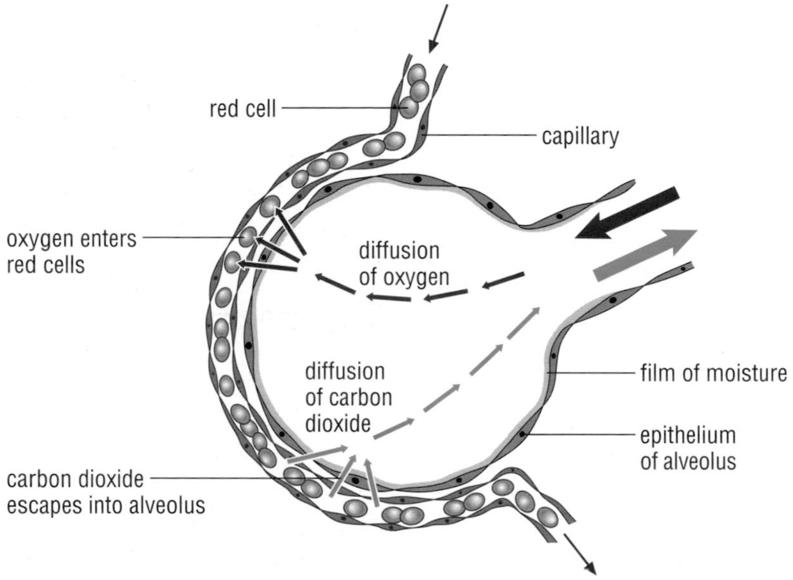

6 How would thick-walled alveoli affect the exchange of the respiratory gases?

7 Compare the way the blood carries oxygen and carbon dioxide.

8 How do you think you would be affected if the surface area of your lungs was reduced?

Figure 6.4   Diagram to show the direction of gaseous exchange.

Moving oxygen to the cells

Once oxygen has entered the red blood cells it begins its journey inside the body. It starts by moving through the capillaries in the lungs. Capillaries are just one of three kinds of blood vessels found in the body (the other two are arteries and veins). They make a network of fine tubes in organs which provide a very large surface area between the blood and the tissues in the organs. This surface area allows a large amount of substances, such as oxygen and glucose, to pass between the blood and the tissues in a short amount of time.

Blood vessels

Arteries

Blood vessels that take blood away from the heart are called arteries. The high pressure of blood pushes strongly on the thick, elastic artery walls. They stretch and recoil as the blood moves by. This movement of the artery wall makes a pulse.

Veins

Blood vessels that bring blood towards the heart are called veins. The blood is not under such high pressure and so does not push as strongly on the vein walls. Veins have thinner walls than arteries and contain valves that stop the blood flowing backwards.

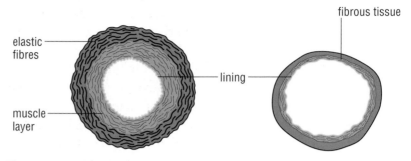

elastic fibres

fibrous tissue

lining

muscle layer

Figure 6.5   An artery (left) and a vein (right) in cross-section.

9 Why do veins have valves?

10 Why do you think arteries function better with thick walls?

Capillaries

When an artery reaches an organ it splits into smaller and smaller vessels. The smallest blood vessels are called capillaries. A capillary wall is only one cell thick. They are spread throughout the organ so that all cells have blood passing close to them. Where the blood leaves an organ, the capillaries join together to form larger and larger vessels until eventually they form veins.

The heart

Movement of the blood is produced by the pumping action of the heart. The heart is divided down the middle into two halves. The right side receives deoxygenated blood from the body and pumps it to the lungs. At the same time the left side receives oxygenated blood from the lungs and pumps it to the body.

The two pumps in the heart and a simplified arrangement of the blood vessels are shown in Figure 6.6.

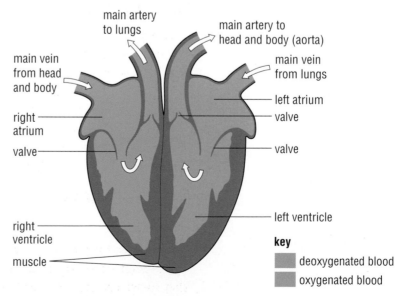

main artery to lungs

main artery to head and body (aorta)

main vein from head and body

main vein from lungs

left atrium

right atrium

valve

valve

valve

left ventricle

right ventricle

muscle

key

deoxygenated blood

oxygenated blood

Figure 6.6   A simplified section through the heart.

11 Where does the pushing force come from to push the blood out of the heart?

12 What is the purpose of the heart valves?

13 Why do you think the walls of the left ventricle are thicker than the walls of the right ventricle?

14 Make a simple drawing of a heart. Above the heart make a simple drawing of the lungs. Below the heart make a simple drawing of a muscle. Now, using the information about how the respiratory gases travel, draw in blood vessels connecting the three organs.

Each side of the heart has two chambers. Blood flows from the veins into the upper chambers called the atria (singular: atrium). It passes from the atria into the lower chambers, the ventricles. The muscular walls of the ventricles relax as they fill up. When these muscular walls contract the blood is pumped into the arteries. Valves between the atria and the ventricles stop the blood going backwards into the atria. Valves between the arteries and the ventricles stop the blood from flowing backwards after it has been pumped out of the heart.

When the blood from the lungs leaves the heart it enters the aorta. This branches to other arteries which connect to different organs of the body. When the blood reaches an organ, it travels into another network of capillaries. It is here that the oxygen leaves the blood, passes through the wall of the capillary and enters the cells where respiration takes place.

Moving glucose to the cells

The energy of glucose is released during respiration. Glucose passes through the wall of the small intestine and into the plasma in the capillaries. It travels in the plasma along veins to the heart. It enters the right side of the heart and then passes to the lungs, where the blood picks up oxygen. It then passes through the left side of the heart and into the aorta. From here it travels along arteries which take the glucose to the body organs.

Moving carbon dioxide to the lungs

Carbon dioxide is a product of aerobic respiration. It leaves the cells where it is produced and passes through the walls of the capillaries and into the plasma. It travels along veins which take it back to the right side of the heart. From here it travels along the main artery to the lungs. It escapes through the capillary walls into the air in the alveoli.

Testing for carbon dioxide

In exhaled breath

Exhaled air can be tested for carbon dioxide by passing it through lime water, Figure 6.7.

For discussion

How could you adapt the apparatus shown in Figure 6.7 to find out if

a) other animals produce carbon dioxide, and if

b) plants produce carbon dioxide?

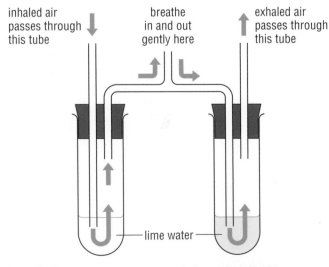

inhaled air passes through this tube

breathe in and out gently here

exhaled air passes through this tube

lime water

Figure 6.7   Testing inhaled and exhaled air for carbon dioxide.

If carbon dioxide is present it reacts with the calcium hydroxide dissolved in the water to produce insoluble calcium carbonate. This makes the lime water turn white or milky.

In air around seeds

Carbon dioxide production can be used as an indication of respiration and a sign of life. Hydrogencarbonate indicator is a liquid that changes colour in the presence of carbon dioxide. It changes from an orange-red colour to yellow. The production of carbon dioxide by germinating pea seeds can be shown by setting up the apparatus shown in Figure 6.8.

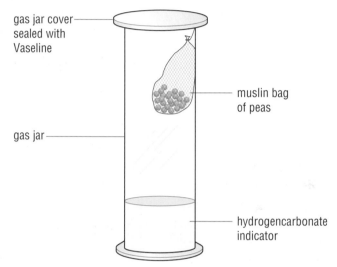

Figure 6.8 Investigating carbon dioxide production by germinating pea seeds.

Summary

- Energy is released from food during respiration (*see page 55*).
- Oxygen and carbon dioxide are exchanged in the respiratory system (*see page 56*).
- Breathing is the process of air exchange between the air and the lungs (*see page 56*).
- Oxygen travels through the blood in the red blood cells (*see page 58*).
- The heart contains two pumps for moving the blood (*see page 59*).
- Glucose travels through the blood in the plasma (*see page 60*).
- Carbon dioxide travels through the blood in the plasma (*see page 60*).
- Carbon dioxide can be detected by indicators (*see pages 60–61*).

End of chapter question

1 Four people took their pulses (measured in beats per minute, on a portable heart monitor) at rest, straight after exercise, one minute after exercise, two minutes after exercise and three minutes after exercise. Here are their results.

Andrew 71, 110, 90, 79, 71
Brenda 74, 115, 89, 77, 73
Clare 73, 125, 115, 108, 91
David 53, 80, 71, 46, 64

a) Make a table of the results.
b) Plot a graph of the results.
c) What trend can you see in the results?
d) When did Andrew's and Brenda's hearts beat at the same rate?
e) Andrew claims to be fitter than Clare. Do you think the results support his claim? Explain your answer.
f) Which result does not follow the trend? Explain why this may be so.

For discussion

Giorgio Baglivi (1668–1707) was an Italian doctor who believed that the body is just a machine. He matched scissors to teeth, bones to levers and lungs to bellows.

a) If he had been alive today, what might he have matched the brain to?

b) Was he right to think of the body as just a machine?

7 Microbes

A microbe is a very tiny living organism, or microorganism. A microscope is needed to see it. The word 'microbe' does not refer to a group of living things in the same way that 'mammal' or 'bird' does. Microbes belong to three kingdoms (see page 37). They are the fungi, the Monera and the Protoctista.

Fungi

Fungi are placed in a separate kingdom because, unlike plants, the cell wall of a fungus contains a substance called chitin. Many kinds of fungi can form very thin threads called hyphae. The hyphae feed on the dead bodies of plants and animals. When the network of hyphae reaches a certain size it forms structures, called sporangia, that release spores. A spore is a reproductive cell with a thick wall around it. The wall provides protection from changes in temperature; it prevents the cell from losing water in dry conditions and from gaining water in wet conditions. In the mould that grows on bread, the sporangia are black spheres.

In some fungi the threads join together to make a larger structure above ground called a fruiting body. This may be divided into a stalk and cap as seen in mushrooms and toadstools. There are spore-producing structures called gills in the cap.

A few fungi are parasites (they feed on living things). Athlete's foot is caused by a fungus that feeds on the skin between toes.

Yeasts are fungi that do not produce hyphae. 'Wild' yeasts feed on the sugar that forms on the surfaces of fruits and in the nectar of flowers. From these 'wild' yeasts, special types have been developed for use in baking and making alcoholic drinks.

Yeast

Yeast respires anaerobically (without the need for oxygen) to produce alcohol and carbon dioxide. It is used to make bread and alcoholic drinks.

Figure 7.1   Mould growing on bread.

> **1** Why are fungi not placed in the plant kingdom?

Figure 7.2   Dough rising to produce the spongy texture of bread.

Bread is made by mixing flour, water, yeast and sugar to form dough. Inside the dough, the yeast respires anaerobically and produces bubbles of carbon dioxide that make the dough rise. Lumps of dough are put into baking tins. When bubbles of gas are heated they expand, so the tins are kept in a warm cupboard for about half an hour to allow the dough to rise even more. They are then placed in an oven for baking. The alcohol evaporates in the heat and a loaf, with a spongy texture, is produced.

Yeast and sugar are used as ingredients to make alcoholic drinks, such as wine and beer. The process of producing alcohol for drinks is called fermentation. During this process bubbles of carbon dioxide are also produced. The alcohol produced mixes with the water in the drink. Alcohol is a poison, so fermentation must be regulated to prevent it increasing to concentrations that kill the yeast. If fermentation was allowed to continue until it stopped, a solution of about 14% alcohol would be produced.

Monera

Living things in this kingdom have bodies made from only one cell that has no nucleus. A major subgroup of the Monera is the bacteria.

Bacteria

Bacteria are single-celled organisms that range in size from 0.001 µm–0.5 µm. They can be seen using a light microscope.

Most bacteria feed on the remains of plants and animals or on animal waste. They feed by secreting a digestive juice which breaks down the food substances around them. Then they draw in the digested food substances for use in growth and reproduction.

Some bacteria can photosynthesise (see Chapter 11) or process chemicals such as hydrogen sulphide to obtain energy for life processes.

Bacteria usually reproduce by a process called fission, in which each bacterium divides into two. If they have enough warmth, moisture and food, some bacteria can reproduce by fission once every 20 minutes.

Bacteria can survive as heat-resistant spores for a long time. They break out of the spores when favourable conditions occur.

2 How does yeast make the spongy texture of the bread?

3 Why does the dough rise even more in a warm oven?

4 Why does fermentation not produce a stronger alcoholic solution than about 14%?

5 If a bacterium could divide into two every 20 minutes, how many bacteria would be produced after 8 hours?

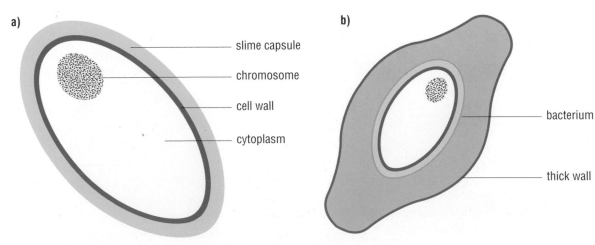

a)
— slime capsule
— chromosome
— cell wall
— cytoplasm

b)
— bacterium
— thick wall

Figure 7.3 a) A bacterium and **b)** a bacterial spore.

Harmful and useful bacteria

Some bacteria feed on the insides of living bodies where they may cause disease. Diphtheria, whooping cough, cholera, typhoid, tuberculosis and food poisoning are all caused by different kinds of bacteria.

Some kinds of bacteria are useful. Yoghurt is made by introducing bacteria to milk and making it turn sour. Vinegar is made by allowing bacteria to feed on ethanol and change it into acetic acid.

Protoctista (single-celled organisms)

These microorganisms may move by making the substance inside their body flow. An *Amoeba* uses this method to form projections called pseudopodia which it uses to catch food. Other members of this kingdom may have a hair-like projection called a flagellum which they lash like a whip to move through water. Many have bodies with small hair-like structures called cilia on their surface. They wave their cilia to-and-fro to push themselves through water.

Protoctista may either take in particles of food or make their own food by photosynthesis. Some live in the bodies of other living things and cause diseases, such as malaria and sleeping sickness.

> **6** In what ways are members of the Monera
> **a)** similar to and
> **b)** different from members of the Protoctista?
> **7** Describe the way Protoctista move.

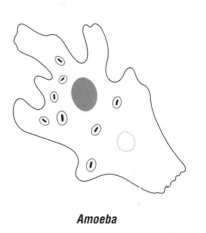

Amoeba

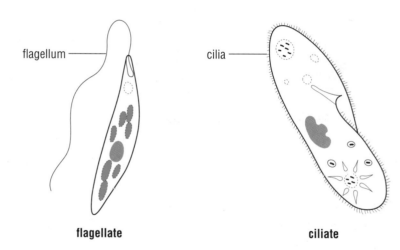

flagellum

cilia

flagellate

ciliate

Figure 7.4 Examples of Protoctista.

Viruses

Viruses are so small that they can be seen only with an electron microscope. Viruses do not show the characteristics of living things such as feeding, respiring or growing. They are usually classified as non-living but they are able to reproduce if they enter a living cell. As they reproduce they destroy the cells they are in. Each kind of virus attacks certain cells in the body. For example, the cold virus attacks the cells that line the inside of the nose. The destructive action of the cold virus on the cells in the nose makes the nose run. In addition to the common cold, viruses can also cause influenza, chicken pox, measles and rabies and can lead to the development of AIDS.

> **8** How are viruses and bacteria different?
> **9** Produce a table of diseases caused by viruses and bacteria.

Transmitting disease

Disease-causing microbes and viruses can be transmitted in several ways, listed overleaf.

Droplets

When people cough or sneeze, they produce a cloud of tiny water droplets which can carry microbes. If other people inhale the droplets they can become infected with the microbes.

Contact

Some diseases, such as leprosy, are spread by an infected person touching an uninfected person.

Water

Water supplies may be contaminated with sewage that contains disease-causing microbes. Cholera is caused in this way.

Food

If food is not prepared in hygienic conditions, harmful bacteria can enter the body, breed and cause food poisoning.

Wounds

The skin provides a protective barrier against microbes, but if it is damaged harmful microbes can enter the tissue below, and can possibly be carried around the body in the blood.

Vectors

Some animals spread disease, especially if they bite a person. For example, the mosquito spreads malaria when it feeds on human blood. The housefly spreads disease by contaminating food (see opposite).

Public health

In developed countries there are public health services that help to keep the population healthy. Two major features of the public health service are water treatment and waste disposal.

Water treatment

Cleaning water to make it fit to drink and treating waste water and sewage are expensive. These processes are found widely only in the wealthy developed countries. In many areas of developing countries there is not enough money available to provide these services and

For discussion

Imagine that you were making an expedition through a rainforest. What are the dangers of there being a disease transmitted to you? What preparations would you make to prevent a disease being transmitted to you?

Figure 7.5 This pump is used to draw up water that has sunk into the ground and settled above water-resistant rocks.

therefore people still suffer from diseases caused by microbes in the water. International aid programmes have been set up to help improve the water supplies in developing countries in order to reduce disease.

Dangerous rubbish

Microbes thrive in decaying household waste such as kitchen scraps. Flies also breed in the waste and carry the microbes on their bodies when they leave the rubbish. If the fly lands on food left out in the kitchen, microbes may be left behind to feed and breed. If the food is then eaten, the microbes may cause illness. This can be prevented by storing rubbish in bins with secure lids and storing food in containers.

Personal hygiene

During the course of the day, sweat wells up from the pores in the skin. The water in the sweat evaporates to cool the body but other substances, such as urea, are left behind. Dirt sticks to the skin and dead skin cells flake off and join the dirt. Bacteria from the dirt or from the air feed and breed on these substances on the skin surface. Their activities make the skin smell and, if the skin is cut, they can enter the body and cause disease. Body odours caused by bacteria on the skin and the risk of infection are greatly reduced by a high standard of personal hygiene. This involves regular, thorough washing of skin and hair, regular changing and washing of clothes, particularly underclothes which are next to the skin, and regular tooth-brushing.

Acne

Hair grows out of tubes in the skin called follicles. There is a sebaceous gland in each follicle that secretes sebum. This substance prevents the skin becoming dry and gives it a waterproof coat. At puberty the body increases the production of sex hormones which may cause an increase in sebum production. The sebum can block the hair follicle and form a blackhead. Bacteria close by may breed and cause the skin to become inflamed to form a spot or pimple. Large numbers of pimples are called acne. In severe acne parts of the sebaceous glands are destroyed and cavities form in the skin. These can leave a scar.

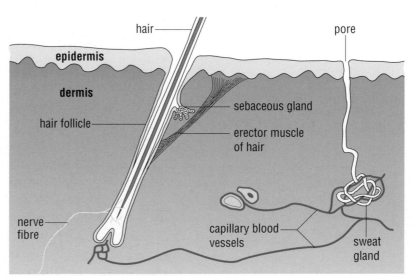

Figure 7.6 Hair follicle in the skin.

10 Why should you clean your skin?

Microbes on teeth

If teeth are not cleaned regularly a sticky layer builds up on the enamel. This layer is called plaque, and bacteria will settle in it. They feed on the sugar in food and make an acid. This breaks down the enamel and creates a cavity that may extend through the dentine into the sensitive pulp cavity.

How the body fights harmful microorganisms

The body's immune system acts to destroy harmful microorganisms. Most microorganisms are killed by white blood cells as they try to enter or soon after they enter the body through a cut, before they do any harm. These white blood cells are a type known as phagocytes.

Microorganisms that do enter the blood and move round the body come into contact with another type of white blood cell, known as lymphocytes. Microorganisms have chemicals called antigens on the surface of their bodies. Lymphocytes detect the antigens and make antibodies to attack the microorganisms and begin to destroy them. The destruction is completed by phagocytes that engulf the attacked microorganisms.

Each kind of microorganism has its own antigen which is different from any other. When the lymphocytes make an antibody to destroy an antigen it can attack only the particular kind of microorganism that makes that antigen. For example, the antibody that helps destroy the bacterium that causes whooping cough will not destroy the bacterium that causes diphtheria.

After a particular kind of disease-causing microorganism has been destroyed, the antibodies remain in the blood for some time. If reinfection occurs, the antibodies can destroy the invading microorganisms quickly before they build up into large numbers to cause disease. Even when the antibodies have left the blood the lymphocytes are quick to detect the antigens of the re-invading microorganism and make antibodies rapidly to begin the destruction process. The action of the lymphocytes gives the body immunity to the disease if the microorganism should reinfect it. In the past, all the immunity that a person had was built up in this way. Today, we are made immune from many microorganisms by vaccination.

Vaccination

The body can be made immune from infection without it having to suffer the full effects of the disease. This is done by vaccination. The body is injected with material that stimulates the lymphocytes to produce antibodies ready to attack those particular microorganisms if they infect the body in the future.

While the immunity given by some vaccines can last a lifetime, other vaccines, such as those for cholera and typhoid, only give immunity for a certain length of time. Extra vaccinations, called boosters, are needed to keep the body immune from these diseases.

For discussion
What should you do to try to keep your teeth for life?

11 What is the difference between
    a) an antibody and an antigen,
    b) a lymphocyte and a phagocyte?
12 How does the body fight reinfection
    a) shortly after it has recovered from the disease,
    b) a long time after it has recovered from the disease?

For discussion
How does society benefit from mass vaccinations against polio?

Passive immunity

The fetus developing in the womb does not make antibodies but receives some from the mother's blood through the placenta (see page 22). After the baby is born it may take in more antibodies through its mother's milk. Immunity built up in this way is called passive immunity. Soon the baby starts to make its own antibodies to build up immunity.

Fighting microorganisms with medicines

Bacteria can be destroyed with an antibiotic. Antibiotics are chemicals that are swallowed or injected to fight microorganisms inside the body. The antibiotic may stop the bacteria from making cell walls or it may affect the life processes taking place inside them. Two well known antibiotics are penicillin and tetracycline.

Viruses cannot be destroyed by antibiotics but some can be destroyed by antiviral drugs. A virus needs substances from the cell it has infected in order to reproduce. An antiviral drug stops the virus reaching these substances.

Antiseptics are chemicals used to attack microorganisms on damaged skin and in the lining of the mouth. They are not swallowed. They are used as creams and mouthwashes.

Athlete's foot is a skin disease caused by a fungus. The fungus feeds on the damp skin between the toes of poorly dried feet. The disease can be cured by treating the feet with powder and cream containing a fungicide which kills the fungus.

13 What are the differences between an antibiotic and an antiviral drug?

Summary

- Microbes are microorganisms that belong to one of three kingdoms of living things – fungi, Monera and Protoctista (*see page 63*).
- Yeast is a useful fungus (*see page 63*).
- Bacteria may be harmless or may cause disease (*see page 65*).
- Viruses are usually classified as non-living (*see page 65*).
- Diseases can be transmitted in several ways (*see pages 65–66*).
- Water treatment is important for community health (*see page 66*).
- Skin hygiene and dental care are important components of personal hygiene (*see pages 66 and 67*).
- The body's immune system naturally helps destroy bacteria and viruses (*see page 68*).
- Immunity can be acquired by vaccination (*see page 68*).
- Medicines can be used to destroy microorganisms (*see above*).

End of chapter questions

1 Which organisms
  a) have cell walls made of chitin,
  b) can make food by photosynthesis,
  c) reproduce by fission?

2 Explain why antibiotics cannot be used to cure the common cold.

3 How are microbes useful?

Ecological relationships

Investigating a habitat

A habitat, such as wood or a pond, is a home for a range of living things. When ecologists study a habitat they need to collect data about it. This not only provides information about the organisms that live there, but the data can also be stored and used to monitor the habitat in the following way. At a later date another survey can be made and the data obtained then can be compared with the previous data. This shows how the populations of species have fared in the time between the surveys. Some may have increased, others decreased, and some stayed the same. By comparing data in this way the effects of events close by, such as a change in land use or the release of pollutants, can be carefully studied for signs of environmental damage.

Recording the plant life in a habitat

When a habitat is chosen for study a map is made in which the habitat boundaries and major features, such as roads or cliffs, are recorded. The main kinds of plants growing in the habitat are identified and the way they are distributed in the habitat is recorded on the map.

A more detailed study of the way the plants are distributed is made by using a quadrat and by making a transect.

Using a quadrat

A quadrat is a square frame. It is placed over an area of ground and the plants inside the frame are recorded. If there are only a few plants in the quadrat, as may occur if it is placed on waste ground, the positions of the individual plants can be recorded in a diagram of the quadrat. If the quadrat is placed in an area with a large number of plants covering the ground, such as in a lawn, the area occupied by each type of plant is estimated. For example, a quadrat on a lawn may show the plants to be 90% grass, 7% daisy and 3% dandelion.

Figure 8.1 Ecologists mapping a habitat.

When using a quadrat the area of ground should not be chosen carefully. A carefully selected area might not give a fair record of the plant life in the habitat but may support an idea that the ecologists have worked out beforehand. To make the test fair the quadrat is thrown over the shoulder so that it will land at random. The plants inside it are then recorded. This method is repeated a number of times and the results of the random samples are used to build up a record of how the plants are distributed. An estimate of how many of each kind there are in the area can then be made.

1 How could you use a quadrat to see how the plants change in a particular area over a year?

Figure 8.2   Ecologists using a quadrat.

Making a transect

If there is a feature such as a bank, a footpath or a hedge in the habitat, the way it affects plant life is investigated by using a line transect. The position of the transect is chosen carefully so that it cuts across the feature being examined.

The transect is made by stretching a length of rope along the line to be examined and recording the plants growing at certain intervals (called stations) along the rope. When plants are being recorded along a

Figure 8.3   Ecologists making a transect.

transect, abiotic factors such as temperature or dampness of the soil may also be recorded to see if there is a pattern between the way the plants are distributed and the varying abiotic factors.

2 How useful do you think quadrats and line transects are for recording the positions of animals?

3 Which method – quadrat or transect – would be more useful for investigating the plants growing around the water's edge of a pond? Explain your answer.

4 Look at the results from a line transect shown in Table 8.1.

Table 8.1

Station	1	2	3	4	5	6	7	8	9	10
Soil condition	W	D	W	W	D	Dr	Dr	Dr	D	Dr
Plant present	A	A	B	A	B	C	C	C	B	C

W = very wet soil; D = damp soil; Dr = dry soil

What can you say about the plants in this habitat from the information in these results?

Collecting small animals

Different species of small animals live in different parts of a habitat. In a land habitat they can be found in the soil, on the soil surface and leaf litter, among the leaf blades and flower stalks of herbaceous plants, and on the branches, twigs and leaves of woody plants. They can be collected from each of these regions using special techniques.

Collecting from soil and leaf litter

A Tullgren funnel is used to collect small animals from a sample of soil or leaf litter. The sample is placed on a gauze above the funnel and a beaker of water is placed below the funnel. The lamp is lowered over the sample and switched on. The heat from the lamp dries the soil or leaf litter and the animals move downwards to the more moist regions below. Finally, the animals move out of the sample and into the funnel. The sides of the funnel are smooth so the animals cannot grip onto them and they fall into the water.

Pitfall trap

The pitfall trap is used to collect small animals that move over the surface of the ground. A hole is dug in the soil to hold two containers, such as yoghurt pots, arranged one inside the other. The containers are placed in the

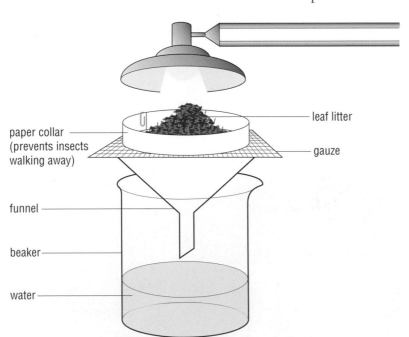

paper collar (prevents insects walking away)

leaf litter

gauze

funnel

beaker

water

Figure 8.4 A simple Tullgren funnel.

5 What is an advantage of using an outer and an inner container instead of just one container for the pitfall trap?

6 Why are large leaves not used inside the trap?

hole, and the gap around them up to the rim of the outer container is filled in with soil. A few small leaves are placed in the bottom of the container and four pebbles are placed in a square around the top of the trap. A piece of wood is put over the trap, resting on the pebbles. The wood makes a roof to keep the rain out and hides the container from predators. When a small animal falls in it cannot climb the smooth walls of the inner container and remains in the trap, hiding under the leaves until the trap is emptied. Traps must be emptied after a few hours and those set in the evening must be emptied the following morning. The animals are collected by removing the cover, taking up the inner container and emptying it into a white enamel dish. The animals can be seen clearly against the white background and identified.

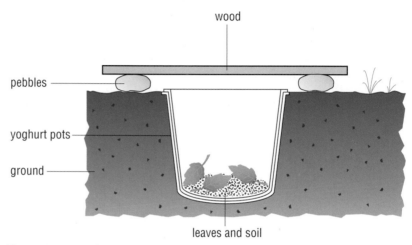

wood

pebbles

yoghurt pots

ground

leaves and soil

Figure 8.5 A pitfall trap.

Sweep net

The sweep net is used to collect small animals from the leaves and flower stems of herbaceous plants, especially grasses. The lower edge of the net should be held slightly forward of the upper edge to scoop up the animals as the net is swept through the plants. After one or two sweeps the mouth of the net should be closed by hand and the contents emptied into a large plastic jar where the animals can be identified.

Figure 8.6 Using a sweep net.

Sheet and beater

Small animals in a bush or tree can be collected by setting a sheet below the branches and then shaking or beating the branches with a stick. The vibrations dislodge the animals, which then fall onto the sheet. The smallest animals can be collected in a pooter (Figure 8.7).

Pooter

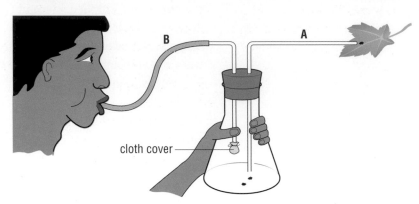

Figure 8.7   Using a pooter.

7 What is the purpose of the cloth on tube B of the pooter?

Tube A of the pooter is placed close to the animal and air is sucked out of tube B. This creates low air pressure in the pooter so that air rushes in through tube A, carrying the small animal with it.

Collecting pond animals

Pond animals may be collected from the bottom of the pond, the water plants around the edges or the open water just below the surface. A drag net is used to collect animals from the bottom of the pond. The net is dragged across the bottom of the pond. As it moves along it scoops up animals living on the surface of the mud. The pond-dipping net is used to sweep through vegetation around the edge of the pond to collect animals living on the leaves and stems. A plankton net is pulled through the open water to collect small animals swimming there.

The drag and pond-dipping nets are emptied into white enamel dishes so that the animals can be identified and studied. The dishes should be

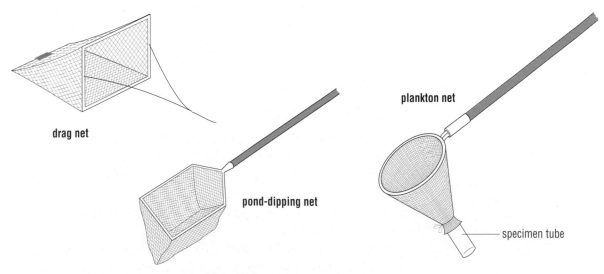

Figure 8.8   Three types of pond net.

set up out of full sunlight so that the water does not get hot, and they should be emptied back into the pond as soon as the investigation is complete. The plankton net has a small bottle that is examined with a hand lens. Samples are then taken for microscope examination.

Studying animals more closely

After a habitat has been surveyed and the different species of plants and animals identified and recorded, further studies about their lives can be made. These involve making observations. If the species to be studied is a plant, a record of its growth, flowering and seed dispersal may be made. If the species is an animal, its behaviour may be observed over different times of the day and throughout the year. Birds are relatively easy to study because most species do not hide away. Small animals hide away but they can be kept in containers in the laboratory for their activities to be observed. Any observations made in the laboratory must be compared to observations of animals living in their habitat before any firm conclusions are drawn, and the laboratory animals should be returned to their habitat as soon as all the observations on them are complete.

Investigating behaviour

Small animals can be studied in the laboratory by setting up a habitat similar to their own. Experiments can be devised to investigate the way the animals behave. For example, it may be noticed that woodlice in a tank with soil, damp moss and pieces of bark are found under the bark in the daytime. This observation may lead to the idea that woodlice do not like the light. This idea can be tested by putting the woodlice in a shallow tray, part of which is uncovered and in the light and part of which is covered and in the dark. The woodlice should be placed in the centre of the tray and left for a few minutes before recording where they have settled to rest.

Ecological pyramids

When a survey of a habitat is complete, ecologists can examine the relationships between the different species. They may find for example that deer depend on ferns for shelter in a wood, and that some birds use moss to line their nests. The major relationship between organisms in a habitat is the relationship through feeding. It is this relationship which interests ecologists (scientists who study ecosystems).

By studying the diets of animals in a habitat, ecologists can work out food chains and ecological pyramids.

Pyramid of numbers

The simplest type of ecological pyramid is the pyramid of numbers. The number of each species in the food chain in a habitat is estimated. The number of plants may be estimated using a quadrat. The number of small animals may be estimated by using traps, nets and beating branches. The number of larger animals, such as birds, may be found by observing and counting.

An ecological pyramid is divided into tiers. There is one tier for each species in the food chain. The bottom tier is used to display information about the plant species or producer. The second tier is used for the

8 Why do you think bird-watching is a popular hobby?
9 Why should laboratory observations be compared with observations of animals in their habitat?

10 Woodlice are generally observed in damp conditions but not in dry areas. How would you set up the shallow tray to test this observation?
11 When checking the behaviour of woodlice in their habitat, some were found under a log one day and under a stone about a metre away the next day. When do you think they moved? Explain your answer.

primary consumer and the tiers above are used for other consumers in the food chain. The size of the tier represents the number of each species in the habitat. If the food chain:

grass → rabbit → fox

is represented as a pyramid of numbers, it will take the form shown in Figure 8.9.

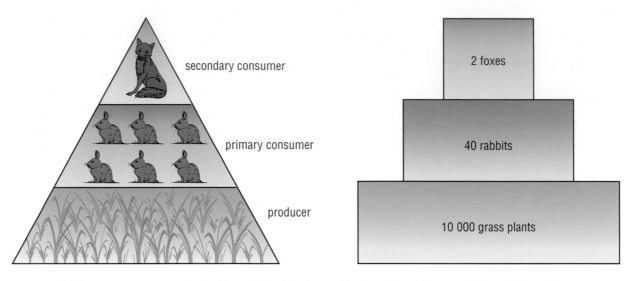

Figure 8.9   Food pyramid of numbers of grass plants, rabbits and foxes.

Not all pyramids of numbers are widest at the base. For example, a tree creeper is a small brown bird with a narrow beak that feeds on insects that in turn feed on an oak tree. The food chain of this feeding relationship is:

oak tree → insects → tree creeper

When the food chain is studied further and a pyramid of numbers is displayed it appears as shown in Figure 8.10.

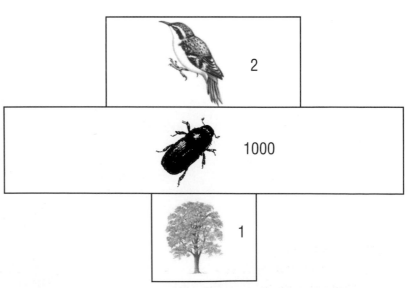

Figure 8.10   Pyramid of numbers of oak tree, insects and tree creepers.

12 What would happen to the numbers of rabbits and grass plants if the number of foxes
   a) increased,
   b) decreased?
13 What would happen to the numbers of grass plants and foxes if the number of rabbits
   a) increased,
   b) decreased?
14 Why do the two food chains considered here produce different pyramids of numbers?
15 Why do you think there are usually more organisms at the bottom of a food chain?

Decomposers

Figure 8.11 These Malaysian termites are feeding on leaf litter.

16 Why are decomposers important? How do they affect you?

The waste of animals and the dead bodies of plants and animals are food for fungi, bacteria and small invertebrates that live in the soil and leaf litter. These organisms are called decomposers. They break down waste and dead remains, and by doing so they recycle the substances from which living things are made so that they can be used again.

Ecosystems

Decomposers form one of the links between the living things in a community and the non-living environment. Green plants form the second link. When a community of living things, such as those that make up a wood, interact with the non-living environment – the decomposers releasing minerals, carbon dioxide and water into the environment and then plants taking them in again – the living and non-living parts form an ecological system or ecosystem. An ecosystem can be quite small, such as a pond, or as large as a lake or a forest.

Working out how everything interacts is very complicated but is essential to ecologists if they are to understand how each species in the ecosystem survives and how it affects other species and the non-living part of the ecosystem. Figure 8.12 shows how the living and non-living parts of a very simple ecosystem react together.

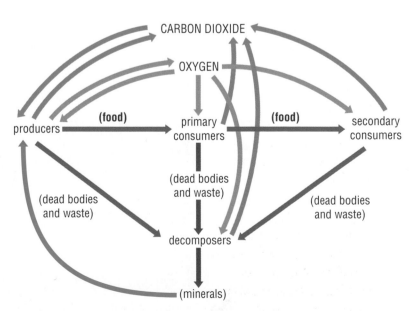

17 How might studying ecosystems help to conserve endangered species?

Figure 8.12 Some relationships in a simple ecosystem.

18 An aquarium tank set up with pond life is an ecosystem. See Figure 8.13.

a) Which are the producers and which are the consumers?

b) Construct some food chains that might occur in the tank.

c) Where are the decomposers?

d) Give examples of the ways the living things react to their non-living environment.

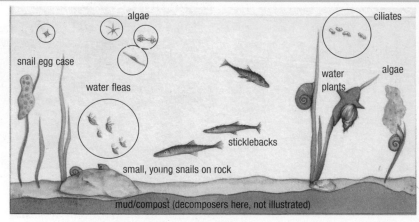

Figure 8.13 An aquarium with pond life. The circled organisms are greatly magnified.

Conservation

Many endangered mammal species have been reduced to a small world population by hunting. The animals have been killed more quickly than they can reproduce. If the death rate exceeds the birth rate the mammal species is set on a course for extinction. Many mammals are now threatened with this course.

They can be helped by raising their birth rate and reducing their death rate. Zoos can help increase the size of the world population of some endangered animal species. They do this by increasing the birth rate by making all the adult animals in their care healthy enough to breed and by providing extra care in the rearing of the young. Zoos also reduce the death rate by protecting the animals from predation. In many countries reserves have been set up in which endangered animals live naturally but are protected from hunting by humans. This reduces the death rate, which in turn increases the birth rate as more animals survive to reach maturity and breed.

For discussion

Large mammals need large areas of natural habitat to support a large population. With the increasing human population why is it difficult to conserve these large areas? Explain your answer.

Figure 8.14 A Java rhinoceros which is threatened with extinction.

Habitat destruction

The first humans did not destroy habitats. They hunted animals and gathered plants to eat in the same way that a few groups of people in the rainforests still do today. When people discovered how to raise crops and farm animals the size of the human population began to rise slightly as there was more food to eat.

In the past few hundred years the size of the human population has grown rapidly as people become healthier, live longer and have more children. Towns and cities have been set up and habitats destroyed to provide the space for them and the farm land needed to support them. Today, with a human population of more than 6 billion, habitats are being destroyed every moment of the day to provide extra room. Tropical rainforests are being destroyed at the rate of an area the size of a football pitch every second to make room for farms, roads and towns.

Figure 8.15 Habitat destruction caused by building a road.

Woodlands are being destroyed to make space for roads, houses and factories. Hedgerows have been removed to make larger fields for growing crops. Freshwater habitats such as lakes and rivers have been polluted with sewage and industrial wastes. These pollutants have also been released into the sea, where marine habitats are also at risk from oil pollution from shipwrecked tankers (see Figure 8.16, overleaf).

Power stations provide electricity. This makes our lives easier but acid rain caused by power station smoke has destroyed forests and polluted lakes and rivers. The carbon dioxide produced by burning coal and oil in power stations adds to the greenhouse effect and leads to global warming. Carbon dioxide in the atmosphere acts like glass in a greenhouse. It lets heat energy from the Sun pass through it to the Earth's surface but it does not let heat from the Earth's surface pass through the atmosphere into space. The heat is trapped in the atmosphere and causes the temperature of the atmospherc to rise.

Figure 8.16 An oil-polluted beach.

Figure 8.17 A freshwater habitat polluted by fertiliser.

---

19 Why do habitats have to be cleared?

20 How have habitats been destroyed by other forms of human activity?

21 Try to imagine the lifestyle of the early humans who hunted animals and gathered plants. Compare this lifestyle with your own. What would you change in your own lifestyle to prevent habitat destruction?

Without this rise in temperature the Earth would be too cold for living things to survive.

Adding more and more carbon dioxide to the atmosphere leads to an increased temperature. This is called global warming and leads to melting of the polar ice, raising of the sea-level and changes in climate in all places on the planet.

Power stations that use nuclear energy do not produce gases that add to the greenhouse effect and global warming. However, they do present the risk of the leak of radioactive materials. The radiation from these materials can destroy any kind of habitat.

Fertilisers increase the amount of food that can be grown and so make it cheaper (see page 116). The use of too much fertiliser has meant that some of it drains into freshwater habitats and can cause an overgrowth of algae. When the algae die back large numbers of bacteria develop to decompose them. The bacteria use up so much oxygen from the water that other forms of water life suffocate and die.

More food, more living space, more electrical energy, more fuel for vehicles and more materials to use in a wide range of ways have all improved the quality of life for millions of people, but to provide all of these things habitats have had to be removed. In the past, when the human population was small, the scale of habitat destruction was also small. Today, with a huge human population, the scale of habitat destruction is vast. Many people feel that more care should be taken in balancing the needs of humans with the destruction of the remaining natural habitats. In many countries there are laws that protect some habitats from destruction and therefore any changes to the remaining habitats have to be carefully planned.

Summary

- Plant distribution in a habitat can be examined by using a quadrat and by making a transect (*see pages 70–71*).
- The Tullgren funnel, pitfall trap, sweep net, sheet and beater can be used to investigate small animals in land habitats (*see pages 72–74*).
- A range of nets may be used to collect small animals from aquatic habitats (*see page 74*).
- Some small animals can be kept in the laboratory and their behaviour can be studied by harmless experiments (*see page 75*).
- A pyramid of numbers shows the numbers of each species at each link in the food chain (*see pages 75–76*).
- Decomposers break down the dead bodies and wastes of living things into simple substances (*see page 77*).
- The living and non-living parts of a habitat form an ecosystem (*see page 77*).
- Humans have had a considerable impact on natural ecosystems and there is a need for conservation (*see pages 78–80*).

End of chapter questions

1 How would you set about investigating a habitat such as a hedgerow?

2 What surfaces do river limpets prefer? The river limpet lives in fast-flowing streams and rivers. There are many different surfaces on the bottom of a stream or river. This experiment was devised to test the results of observations in streams and rivers. The experiment also tested limpets of different sizes to see if larger ones had different preferences to smaller ones.

A 9 cm crystallising dish was divided into four sections called sectors. Each sector had one type of material, either sand, grit, stone or glass. The sectors were covered with water. Twelve limpets were used in each trial and there were two trials for each size class. New limpets were used for each trial. During each trial the number of limpets in each section was recorded after 30 minutes and 60 minutes. The results are displayed in Table 8.2.

Table 8.2

Size class (mm)	30 minutes				Total		60 minutes				Total	
	Sand	Grit	Stone	Glass	Rough	Smooth	Sand	Grit	Stone	Glass	Rough	Smooth
3–4	5	1	9	9	6	18	3	0	15	6	3	21
4–5	1	1	17	5	2	22	0	0	19	5	0	24
5–6	1	1	17	5	2	22	0	0	20	4	0	24
6–7	2	3	11	8	5	19	0	0	21	3	0	24
Total	9	6	54	27	15	81	3	0	75	18	3	93

(continued)

a) How many
   i)   size classes were there,
   ii) trials were made in total,
   iii) limpets took part in the experiment?
b) After 30 minutes what can you conclude about the surface the limpets preferred?
c) What has happened in the crystallising dish during the period from 30 to 60 minutes?
d) What can you conclude about the limpets' surface preferences?
e) Do different sized limpets have different surface preferences?
f) A limpet has a foot like a snail or slug. Why do you think it prefers the surfaces shown in the results?
g) If the experiment was extended to 2 hours what results would you predict?

3 The tufted hair grass forms a clump called a tussock. This provides a habitat for different kinds of invertebrates. The numbers of individuals of the different kinds of invertebrates were investigated over an 8-month period.

Fifteen tussocks were dug up at random on a common each month and were taken apart carefully. The animals in each tussock were collected, arranged into groups and counted. The graphs in Figure 8.19 (opposite) were produced from the data collected.
a) How did the numbers of butterflies and moths (larvae and pupae) change during the 8 months?
b) Describe the population of animals in the tussocks in April.
c) How did the population change by May?
d) Why were the tussocks picked at random?

The animals found in the tussocks of tufted hair grass were compared with those in the tussocks of cocksfoot grass in May. The results were displayed in a bar chart (Figure 8.18).

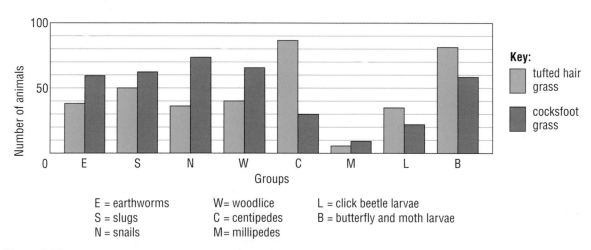

E = earthworms   W= woodlice   L = click beetle larvae
S = slugs   C = centipedes   B = butterfly and moth larvae
N = snails   M= millipedes

Figure 8.18

e) What are the two most numerous groups of animals in
   i)   the tufted hair grass tussocks,
   ii) the cocksfoot tussocks?
f) Which group of animals is found almost in the same numbers in both tussocks?
g) Which group of animals is twice as numerous in one type of tussock compared to the other?

(continued)

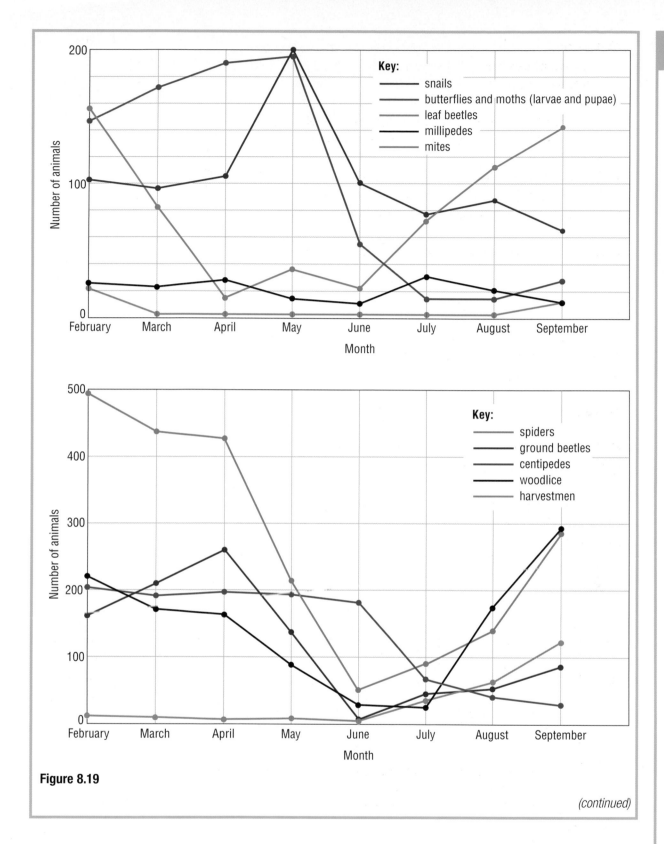

Figure 8.19

(continued)

The animals found in the tussocks of the cocksfoot grass in June were compared with those found in a rush tussock. The two pie charts in Figure 8.20 show how the populations compare.

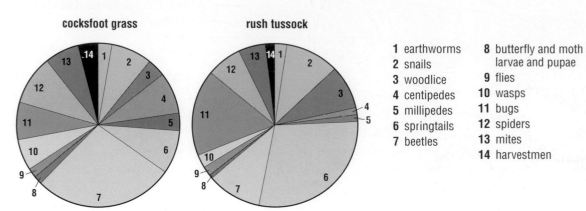

1 earthworms
2 snails
3 woodlice
4 centipedes
5 millipedes
6 springtails
7 beetles

8 butterfly and moth larvae and pupae
9 flies
10 wasps
11 bugs
12 spiders
13 mites
14 harvestmen

Figure 8.20

h) What are the major ways in which the animal populations in the two tussocks differ?

Inheritance and selection

Cell division

Living things are made up of cells. It is the nucleus in a cell that controls the growth and development of the cell. The nucleus also provides the cell with all the instructions to carry out life processes.

Many cells are able to make new cells by dividing into two. This is necessary for organisns to grow. When a cell in a body divides, the nucleus divides first. Each new nucleus forms the nucleus of a daughter cell.

Just before a cell divides, long strands of material appear in the nucleus. These strands are called chromosomes.

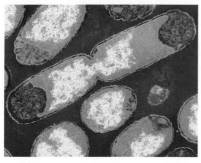

Figure 9.1   Two cells almost divided.

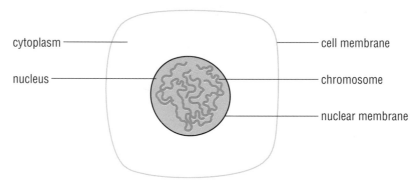

Figure 9.2   Chromosomes in the nucleus of a cell.

As the nucleus begins to divide, each chromosome makes a copy of itself, forming two threads. During division of the nucleus, each thread separates from the other and enters one of the new nuclei. Here it forms a chromosome. This means that the nuclei of the daughter cells have exactly the same number of chromosomes as the nucleus of the parent cell.

The chromosomes in the new nuclei of the daughter cells disappear back into the nuclear materials but reappear again when the cells are about to divide.

Formation of gametes

Gametes, such as eggs and sperm, are the sex cells of an organism. They join together in a process called fertilisation to produce a zygote (see page 21). If the gametes were produced in the same way as ordinary body cells, they would have the same number of chromosomes as body cells. This would create a problem at fertilisation, as the zygote would have twice as many chromosomes as its parents. If zygotes from this generation formed individuals which bred, the zygotes produced would have four times the number of chromosomes as the grandparents. Within a few generations the zygotes produced would be so packed with chromosomes that they would die. This situation does not develop in real life because the gametes are produced by a different kind of cell division.

In a cell of the reproductive organ that is to produce gametes, the chromosomes appear. In every body cell there are pairs of chromosomes. Each one of a pair has a similar appearance to the other (the sex chromosomes are the exception, see opposite) and one chromosome of the pair is originally from the male parent and the other is from the female parent. In the cell division that forms the gametes, only one chromosome of each pair goes into the gamete. Gametes thus contain half the number of chromosomes formed in body cells.

When fertilisation occurs, the nuclei of the gametes, each containing half the number of chromosomes of the normal body cells, join together. The nucleus of the zygote that is formed has the same number of chromosomes as the body cells of the parents. This means that the number of chromosomes in a zygote does not increase with each generation.

Each species has a certain number of chromosomes in the body cells. Human body cells have 46 chromosomes. Figure 9.3 shows how chromosomes pair up at fertilisation, and how an embryo develops with body cells that have the same number of chromosomes as the parent.

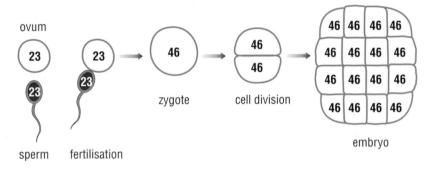

Figure 9.3   How chromosomes pair up at fertilisation.

All four gametes formed by a sperm-producing cell develop into sperm, but only one gamete formed from an ovum-producing cell develops into an egg or ovum.

A closer look at variation

In Chapter 4 we looked at how features vary within a species. Here we will look at how the features relate from one generation to the next. Figure 9.4 shows the faces of the parents and four children in a family. Features seen in the parents are also seen in the children, but the children are not exact copies of their parents.

The features, such as hair colour and shape (wavy or straight), eye colour and the presence or absence of freckles and ear lobes, are present in both generations and have been passed from one generation to the next.

> 1 Look at Figure 9.4. What features have
> a) Ashley,
> b) Bryony,
> c) Charles,
> d) Davina inherited from their mother and father?

Figure 9.4   Variation in a family.

Chromosomes and genes

Before we look at how features are inherited we need to look at how the features are produced. The answer lies in the structure of the chromosomes. Chromosomes are really threads of chemical messages. The messages are strung along the chromosome like carriages in an inter-city train. Each message is called a gene. The genes provide all the information for how a cell grows, develops and behaves. The genes of a chromosome are made of a substance called deoxyribonucleic acid (DNA). Small variations in the structure of DNA produce the different genes in our bodies and different forms of the genes in different people.

How genes pass on

When gametes are made, the genes pass into them on the chromosomes. During the formation of the gametes, parts of the chromosomes may swap portions. This swapping leads to a mixing up of the genes, so an exact copy of the parent's genetic code is not passed on.

When a zygote is formed at fertilisation, the nucleus contains all the genes needed to make a new individual. As there has been some mixing of the genes from both parents, the new individual develops a slightly different combination of features from their parents.

How males and females are produced

One pair of chromosomes are the sex chromosomes. There are two sex chromosomes and in humans they have different lengths. The longer one is called the X chromosome and the shorter one is called the Y chromosome. Females have two X chromosomes and males have an X and a Y chromosome. When the male makes sperm cells each one receives either an X or a Y chromosome as the pair divides. All the eggs receive an X chromosome because female cells do not contain a Y chromosome.

When an egg meets a sperm at fertilisation it has an equal chance of meeting a sperm containing an X or a Y chromosome because the sperms with the different sex chromosomes are produced in equal numbers.

2 What will be the sex of a baby produced when a sperm containing a Y chromosome fertilises an egg? Explain your answer.

3 Why do people not have two Y chromosomes?

4 Are girls or boys more likely to be formed after fertilisation? Explain your answer.

Species and varieties

In Chapter 4 we saw that living organisms are divided into groups called species. All the members of a species have the same number of chromosomes in their body cells. For example dogs have 78 chromosomes while cats have only 38; potatoes have 48 while peas have only 14.

When the males and females in a species reproduce they produce offspring which have the same number of chromosomes as themselves. The offspring are also capable of reproducing.

Most species have means which prevent them from breeding with other species. For example, if the pollen from one plant lands on the stigma of a different species it may not produce pollen tubes because the correct concentration of sugar is not present. The behaviour of cats and dogs prevents them from breeding together, but even if their sperm and eggs were mixed the difference in the number of their chromosomes would prevent them from combining and producing offspring with the same number of chromosomes as their parents.

However, within a species different varieties can be produced. The members of different varieties of a species have the same number of chromosomes as each other but they have different combinations of genes which give a different combination of features. For example, by carefully breeding tomato plants, tomatoes can be produced in red, orange and yellow. They can also be produced as small as cherries or as large as grapefruit to suit the requirements of different people. This is an example of selective breeding.

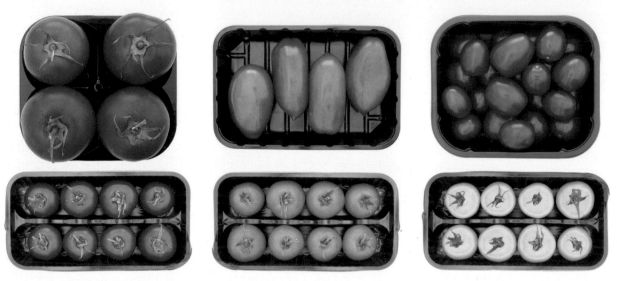

Figure 9.5 Different varieties of tomato.

Selective breeding

For thousands of years people have been breeding animals and plants for special purposes. Most plants were originally bred to produce more food but later plants were also bred for decoration. Animals were originally bred for domestication, then for food production or to pull carts.

A breeding programme involves selecting organisms with the desired features and breeding them together. The variation in the offspring is examined and those with the desired feature are selected for further breeding. For example, the 'wild' form of wheat makes few grains at the top of its stalk. Individuals that produce the most grains are selected for

Figure 9.6 'Wild' wheat (left) and modern wheat (right).

5 All the different breeds of dog have been developed from the wolf by selective breeding. Look at Figure 9.7. What features do you think have been selected to produce a greyhound? Give a reason for each feature you mention.

breeding together. When their offspring are produced they are examined and the highest grain producers are selected and bred together. By following this programme wheat plants producing large numbers of grains have been developed.

In some breeding programmes a number of features are selected and brought together. The large number of different breeds of dog have been developed in this way.

Natural selection

Just as humans select individuals for further breeding, it has been found that selection in a habitat takes place naturally. Those individuals of a species which have the most suitable features to survive will continue to live and breed and pass on their features. Individuals which have features that do not equip them for survival will perish, and so do not pass on their features to future generations. This form of selection is used to explain the theory of evolution, where one species changes in time until another species is produced. The finches on the Galapagos Islands, first studied by Charles Darwin (1809–1882), are thought to have evolved by natural selection (see Figure 9.8).

Figure 9.7   A wolf and a greyhound.

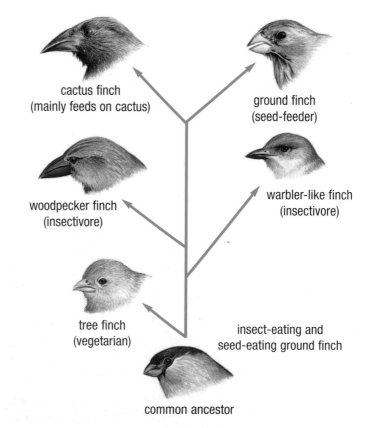

cactus finch
(mainly feeds on cactus)

ground finch
(seed-feeder)

woodpecker finch
(insectivore)

warbler-like finch
(insectivore)

tree finch
(vegetarian)

insect-eating and
seed-eating ground finch

common ancestor

Figure 9.8   Darwin's finches

For discussion

How do you think that the human species may evolve in the future?

INHERITANCE AND SELECTION

Clones

A clone is an organism that is an exact copy of its parent. Clones can occur naturally through a process called asexual reproduction. *Amoeba* reproduces asexually when it simply divides in two. *Hydra*, a small animal related to sea anemones which lives in ponds and ditches, grows a bud which detaches itself and becomes a copy of its parent. Some plants can also form clones. The spider plant is a familiar houseplant which grows small plants on side shoots. The small plants can become detached and live on their own. A simple way of cloning a plant is to take a cutting from it and grow the cutting in compost to make a new plant.

A technique for cloning animals has been developed using nuclei and cells. The nucleus is removed from an egg cell and is replaced by the nucleus of a normal body cell from the animal you wish to clone. The egg is then allowed to develop normally and an exact copy of the animal is produced. Dolly the sheep was the first successfully reared clone at the end of the 20th century. Since then many other clones have been made.

Summary

- The nucleus of a cell contains chromosomes (*see page 85*).
- When a cell divides each new nucleus receives chromosomes from the parent cell (*see page 85*).
- A special type of cell division occurs in the reproductive organs that makes sure the following generation has the same number of chromosomes in their cells as the parent generation (*see page 86*).
- There are genes on chromosomes which carry instructions for the cell (*see page 87*).
- The sex of an individual is determined by the sex chromosomes (*see page 87*).
- There are varieties within species (*see page 87*).
- New varieties of a species can be produced by selective breeding (*see page 88*).
- Natural selection is the process by which new species evolve (*see page 89*).

> **For discussion**
>
> How could cloning help farming?
>
> What disadvantages might there be to cloning farm animals?
>
> If you were cloned today would your clone be just like you when it is your age? If not, why not?

End of chapter questions

1 a) The body cell of a wheat plant has 42 chromosomes. How many chromosomes are in the nucleus of an egg cell?

b) The body cell of a chicken has 78 chromosomes. How many chromosomes are in the nucleus of a sperm cell?

c) The nucleus of a pollen grain of a cabbage has 9 chromosomes. How many chromosomes are in the nucleus of a body cell?

d) The egg of a housefly has 6 chromosomes. How many chromosomes are in the nucleus of a body cell?

2 Camouflage helps animals survive in their habitat through the process of natural selection. Investigate camouflage by studying the results of the following experiment.

(continued)

Garden birds were given different coloured food pellets against different coloured backgrounds to test the effect of camouflage. The food pellets were made of flour and lard and were coloured with harmless dyes. The pellets represented prey animals and were coloured either green or brown. The prey were tested against three backgrounds – grey, green and brown. The backgrounds were painted on 50 cm square aluminium sheets.

Twenty-four trials were made, twelve with grey backgrounds and six each with green and brown backgrounds. In each trial two backgrounds of the same colour were set up a metre apart in a garden. Ten green and ten brown prey were placed on each sheet. When half the prey had been eaten by the garden birds the trial was stopped and the numbers of green and brown prey that had been eaten were recorded. The trials with pairs of coloured backgrounds were made in a random order. The results of the trials are displayed in Table 9.1.

Table 9.1

Grey background				Green background		Brown background	
Numbers eaten		Numbers eaten		Numbers eaten		Numbers eaten	
Green	Brown	Green	Brown	Green	Brown	Green	Brown
18	10	8	19	6	17	11	2
11	8	14	5	0	11	10	2
8	13	7	11	15	4	19	1
12	4	15	6	12	12	5	7
18	9	14	4	12	19	12	8
4	12	18	17	11	9	5	11
Grand totals:		147	118	56	72	62	31

a) Why were two backgrounds used in each trial?
b) Why were the backgrounds mixed up at random and not put out as grey twelve times, green six times and brown six times?
c) Were green or brown prey more likely to be eaten on the grey background?
d) Identify any pairs of numbers that do not follow this trend.
e) What percentage of green prey are eaten on a grey background?
f) What is the percentage of green prey eaten on
   i) a green background,
   ii) a brown background?
g) How might you have expected the results on the green background to be different from those in the table?
h) Do the colours of the prey help to camouflage them? Explain your answer.

Keeping healthy

A body is made up of billions of cells. Groups of similar cells form tissues that work together with other tissues in larger structures called organs. The organs form groups called organ systems (see page 1) which work together to keep the body alive. The power to make the organ systems work comes from the energy in food, which is released by chemical reactions. The way the body is made and the way it works are incredibly complicated. If you live a healthy lifestyle and have a healthy diet (see Chapter 5) your body may keep working for over 80 years. If you live an unhealthy lifestyle the cells, tissues and organs may become damaged.

Figure 10.1 A family enjoying a balanced meal.

Exercise

Regular exercise makes many of the organ systems become more efficient. It also uses up energy and helps to prevent large amounts of fat building up in the body. Exercise can increase your fitness in three ways: it can improve your strength, make your body more flexible and less likely to suffer from sprains, and it can increase your endurance which is your ability to exercise steadily for long periods without resting. Different activities require different levels of fitness. Table 10.1 shows these levels for different sporting activities. By studying the table you can work out which activities you could do to develop one or more of the three components of fitness.

1 Which activities demand great flexibility?
2 Which activity is the least demanding?
3 Which activities are the most demanding?
4 How do the demands of soccer and long distance running compare?
5 Which activity would you choose from the table? What are its strengths and weaknesses?
6 Many people claim that they do not have time to exercise. How would you motivate such people to take some form of exercise? Which activities might suit them best?

Table 10.1

Activity	Strength	Flexibility	Endurance
Basketball	✔✔	✔✔	✔✔✔
Dancing	✔✔	✔✔✔	✔✔
Golf	✔✔	✔✔	✔✔
Long distance running	✔✔✔	✔✔	✔✔✔
Soccer	✔✔	✔✔	✔✔✔
Squash	✔✔✔	✔✔✔	✔✔✔
Swimming	✔✔✔	✔✔	✔✔✔
Tennis	✔✔✔	✔✔✔	✔✔✔
Walking	✔	✔	✔✔

Smoking and health

In Chapter 6 we saw how the respiratory system works to provide us with an exchange of respiratory gases. An efficient exchange is needed for good health. When people smoke they damage their respiratory system and risk seriously damaging their health.

There are over a thousand different chemicals in cigarette smoke, including the highly addictive nicotine. These chemicals swirl around the air passages when a smoker inhales and touch the air passage linings. In a healthy person, dust particles are trapped in mucus and moved up to the throat by the beating of microscopic hairs called cilia. The small amounts of dust and mucus are then swallowed. In a smoker's respiratory system the cilia stop beating due to chemical damage by the smoke. More mucus is produced but instead of being carried up by the cilia it is coughed up by a jet of air as the lungs exhale strongly. This is a smoker's cough and the amount of dirty mucus reaching the throat may be too much to swallow.

In time chronic bronchitis may develop. The lining of the bronchi become inflamed and open to infection from microorganisms. The inflammation of the air passages makes breathing more difficult and the smoker develops a permanent cough. The coughing causes the walls of some of the alveoli in the lungs to burst. When this happens the surface area of the lungs in contact with the air is reduced. This leads to a disease called emphysema.

Some of the cells lining the air passages are killed by the chemicals in the smoke. They are replaced by cells below them as they divide and grow. Some of these cells may be damaged by the smoke too and as they divide they may form cancer cells. These cells replace the normal cells in the tissues around them but they do not perform the functions of the cells they replace. The cancer cells continue to divide and form a lump called a tumour. This may block the airway or break up and spread to other parts of the lung where more tumours can develop.

7 What is the function of a smoker's cough?

8 Why may chronic bronchitis lead to other diseases?

9 How does the reduced number of alveoli affect the exchange of oxygen and carbon dioxide?

10 Why does someone with emphysema breathe more rapidly than a healthy person?

11 How are cancer cells different from normal cells in the lung tissue?

12 Why do cancer cells in an organ make the organ less efficient?

13 Why might the growth of cancer tumours in an organ have fatal results?

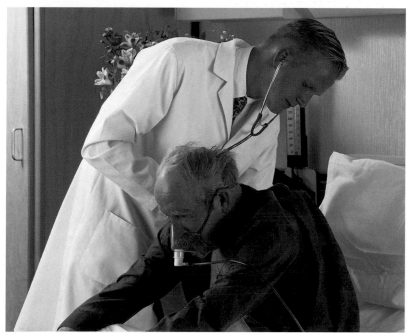

Figure 10.2 Emphysema is one effect of smoking on the respiratory system.

14 How does tobacco smoke contribute to heart and arterial disease?

Carbon monoxide and nicotine in tobacco smoke greatly increase the risk of heart and arterial (cardio-vascular) disease in smokers. Carbon monoxide damages the lining of the arteries; nicotine raises heart rate and blood pressure and increases the risk of blood clotting (thrombosis). These effects help explain the high death rate of smokers from cardio-vascular disease. The risk is further increased in smokers who do not take regular exercise or who have a diet high in fat.

A healthy diet

In Chapter 5 we looked in detail at the importance of diet. From this work you can think about the amounts of food you eat in the following way.

You can eat large amounts of potatoes, bread, rice and pasta. They provide you with carbohydrates which supply the body with energy. You should eat five portions a day of fruit and vegetables to provide you with vitamins, minerals and fibre. You should eat a smaller amount of foods such as meat and fish, to provide you with the protein you need for growth and repair of the body. In a vegetarian diet protein is mainly provided by pulses such as beans, lentils and peas. Finally, you should eat only very small amounts of food rich in fat such as chocolate, nuts, fatty meat and cheese. Fat provides materials for making cell membranes. It also creates an insulating layer beneath the skin that helps to retain body heat. This layer acts as an energy store for the body, but there can be dangers to health if it becomes too thick. Excess body weight, known as obesity, can be caused by eating too much high-energy, particularly high-fat, food. Too much high-fat food also increases the risk of heart disease (see page 96).

An easy way to think about a healthy diet is to imagine a pyramid of food, as Figure 10.3 shows.

The pyramid reminds you that you should eat large amounts of foods at the bottom of the pyramid but only small amounts of those at the top.

15 Make a pyramid of food representing your diet. Describe how it compares with the pyramid in Figure 10.3. If it does not match the pyramid in the diagram, how can you change your diet to make it healthier?
Note: this pyramid of food is simply to make you think about eating healthily. It is not related to ecological pyramids found on page 76.

Figure 10.3 A pyramid of food representing a healthy diet.

Alcohol

Many people have their first alcoholic drink at special family occasions such as birthdays and weddings. As they get older they may have a few drinks at weekends then perhaps during the week too. Drinking alcohol may be used at first as a form of relaxation with friends but may become a regular daily habit until the person cannot cope without drinking large amounts daily. This final stage is alcoholism. To avoid this people need to develop a sensible attitude towards alcohol.

Effect of alcoholic drinks

Alcohol affects the nervous system. It slows down the speed at which nerve cells carry signals. A small amount of alcohol may make a person feel more relaxed but it also makes the nerves work slightly more slowly. This makes a person react more slowly. As the person drinks more, the effect of the alcohol on the nervous system increases and their reaction time becomes longer. The behaviour of the person may change. Their voice may become louder and they may become reckless and even aggressive. The person finds it more difficult to think and speak clearly. If they continue to drink, their body movements may lose co-ordination and the person may be unable to walk. They may fall asleep or become unconscious. In extreme cases, in the unconscious state they may be sick and if the vomit gets stuck in their windpipe they may suffocate and die.

Long-term effects of alcohol abuse

Alcohol is a poison. The liver collects poisons from the blood as it flows through. It breaks down the poisons to make them harmless. If large amounts of alcohol are drunk over many years the liver may become inflamed and develop a disease called hepatitis. Parts of the liver may turn to scar tissue. This leads to the development of cirrhosis of the liver which reduces the liver's capacity to neutralise poisons. This disorder can be fatal.

Strength of alcoholic drinks

Bottles and cans of alcoholic drinks have the strength of their alcohol content marked on them. It is shown as a number with % ABV written after it. This means % of alcohol by volume. Many beers have a strength of 3.5% ABV, while cans of lager or cider may be much stronger, having a strength of 9% ABV. Whisky may be 40% ABV, while sherry is about 15% ABV and wine about 12% ABV. Some 'alcopops' may be up to 13.5% ABV.

Effect of solvent sniffing

When they are sniffed, the solvents in a range of household products can produce an effect similar to drunkenness. The solvents that are sniffed are usually in certain kinds of glues, correcting fluids and aerosol sprays. The butane gas used as the fuel in cigarette lighters is also inhaled. The effect of a solvent on the body occurs very quickly because the chemicals enter the blood through the thin lining of the lungs. The effect does not last long. If a person wants to stay under the effect of the solvent he/she has to keep sniffing.

People react differently to the chemicals in solvents. Some young people have died after sniffing solvents for the first time. The solvents can damage lungs, the control of breathing, the nervous system, liver, kidneys and bone

16 Arrange wine, beer, 'alcopops', whisky, sherry and cider in order of their alcoholic content, putting the strongest one first.

For discussion
Do you think 'alcopops' are suitable alcoholic drinks for young people?

marrow. As sniffers lose control of themselves, they may suffocate on the plastic bags they use to inhale the solvent, become unconscious then be sick and suffocate in their vomit, or cause a fire and risk burning themselves when lighting a cigarette near solvents that are flammable.

A healthy heart

The human heart starts to form in the embryo 20 days after conception. Fewer than 1% of babies are born with heart defects, yet in the United Kingdom more people die from heart disease than from any other disease.

The heart may beat up to 2500 million times during a person's life. Its function is to push blood around the vast network of blood vessels in the body. This push creates a blood pressure that drives the blood through the blood vessels. As the ventricles in the heart fill with blood the pressure in the blood vessels is reduced, but as the heart pumps it out along the arteries the blood pressure rises. In young people the walls of the arteries are elastic and they stretch and contract with the changing blood pressure. Also, the diameters of the arteries are large enough to let the blood flow with ease. As the body ages the artery walls become less elastic.

The heart has its own blood vessels called the coronary arteries and veins. They transport blood to and from the heart muscle.

Fatty substances, such as cholesterol, stick to the walls of arteries and begin to form a raised patch called an atheroma. The blood then has less space to pass along the arteries and its pressure rises as it pushes through the narrower tubes. The developing atheroma may cause a blood clot which narrows the artery even more or can completely block it, causing a thrombosis. This means that the artery is unable to transport oxygen and other nutrients. A thrombosis in a coronary artery causes a heart attack. A thrombosis in an artery in the brain causes a stroke.

The features that develop in the body that cause heart disease can be inherited. People whose relatives have suffered from heart disease should take special care to keep their heart and circulatory system healthy.

Keeping the heart healthy

The heart is made of muscle and like all muscles it needs exercise if it is to remain strong. The heart muscles are exercised when you take part in the activities in Table 10.1 (see page 92). Heart muscle contracts more quickly and more powerfully during exercise than it does at rest so that more blood can be pumped to your muscles. These muscles need more blood to provide extra oxygen while they work.

As we have seen, the blood supply to heart muscles can be reduced by fatty substances such as cholesterol in the blood. These substances are formed after the digestion of fatty foods. Some fatty substances are needed to keep the membranes of the cells healthy, but too much intake of fat leads to heart disease. A heart can be kept healthy by cutting down on the amount of fat in the diet. This may be achieved by cutting fat off meat, or eating fewer crisps and chips for example.

The avoidance of smoking is also very important in helping to maintain a healthy heart. Smoking is a contributory factor in the build-up of fatty deposits in the blood vessels of the heart itself. This increases the risk of heart disease and, ultimately, of a heart attack.

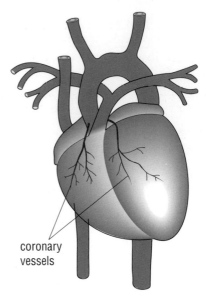

coronary vessels

Figure 10.4 The coronary blood vessels.

The skeleton

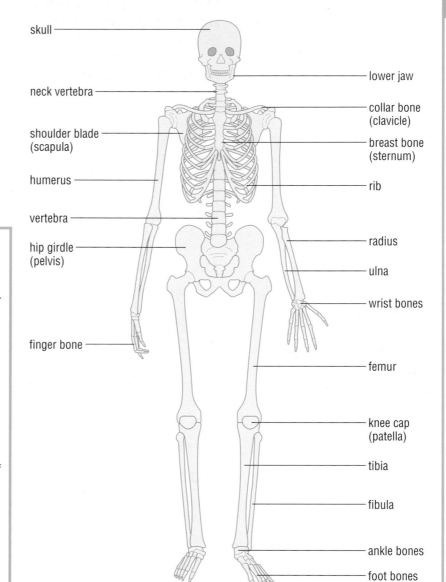

skull

neck vertebra

shoulder blade
(scapula)

humerus

vertebra

hip girdle
(pelvis)

finger bone

lower jaw

collar bone
(clavicle)

breast bone
(sternum)

rib

radius

ulna

wrist bones

femur

knee cap
(patella)

tibia

fibula

ankle bones

foot bones

Figure 10.5 The human skeleton.

17 There are three bones in the arm. How many are in the wrist and hand?

18 How many bones are in all four limbs?

19 A person has a mass of 43.5 kg. What is the mass of their skeleton?

20 Figure 10.5 shows the main bones of the body. How many of these bones can you feel in your body?

21 The skull forms a solid sheet of protection and the ribs form a cage. Why do you think the rib cage is not a solid sheet like the skull? Which offers the better protection, the sheet or the cage? Explain your answer.

22 Newborn mammals have soft skeletons to allow some flexibility during birth. All mammals are fed milk by their mothers. What effect will this have on their bones?

There are 206 bones in the human skeleton. Each arm and hand together have 30 bones. Each leg and foot together have 29 bones. The skeleton accounts for 15% of the mass of the body. The tissue of the skeleton (bone) is hardened as it takes up calcium from the digested food.

The skeleton and protection

The brain and the spinal cord form the central nervous system and are made from soft tissue. The bones of the skull are fused together to make a strong protective case around the brain (see Figure 10.6, overleaf). The backbone is made up of vertebrae (singular: vertebra). There is a hole in each vertebra through which the spinal cord runs. The column of vertebrae makes a protective tube of bone around the spinal cord. There are gaps between the vertebrae through which nerves pass from the spinal cord to the body. The ribs and backbone form a protective structure around the lungs and the heart.

For discussion

What would the body be like without a skeleton?

Could the body survive without a skeleton?

KEEPING HEALTHY

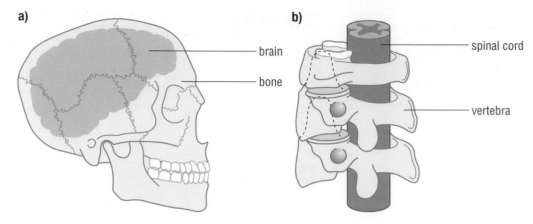

Figure 10.6 Protection of the central nervous system: **a)** the brain and **b)** the spinal cord.

The skeleton and support

The organs that form systems such as digestive, circulatory, excretory and respiratory systems account for 20% of the body's weight. The organs are made from soft material and have no strong supporting material inside them. The bones of the skeleton provide a strong structure to which the organs are attached. They allow the organs to be spread out in the body without squashing into each other.

The muscles account for 45% of the body's weight. They are also made from soft tissue but gain their support from the bones to which they are attached.

The skeleton and movement

The place where bones meet is called a joint. In some joints, such as those in the skull, the bones are fused together and cannot move. Most joints, however, allow some movement. Some joints such as the elbow or knee are called hinge joints, because the movement is like the hinge on a door. The bones can only move forwards or backwards. A few joints, such as the hip, are called ball and socket joints because the end of one bone forms a round structure like a ball which fits into a cup-shaped socket. These joints allow much more movement.

---

23 A person has a mass of 43.5 kg. How much of this mass is due to
   a) their organ systems,
   b) their muscles?

24 The percentage of the body's mass not accounted for by the skeleton, organs and muscles is due to fat. What percentage of the body's mass is due to fat?

25 Look at the skeleton in Figure 10.5 (page 97). Which bones meet at
   a) the hip joint,
   b) the knee joint,
   c) the elbow joint,
   d) the shoulder joint?

26 Name
   a) two hinge joints and
   b) two ball and socket joints.

27 How might a joint be affected by
   a) torn ligaments,
   b) lack of synovial fluid,
   c) damaged cartilage?

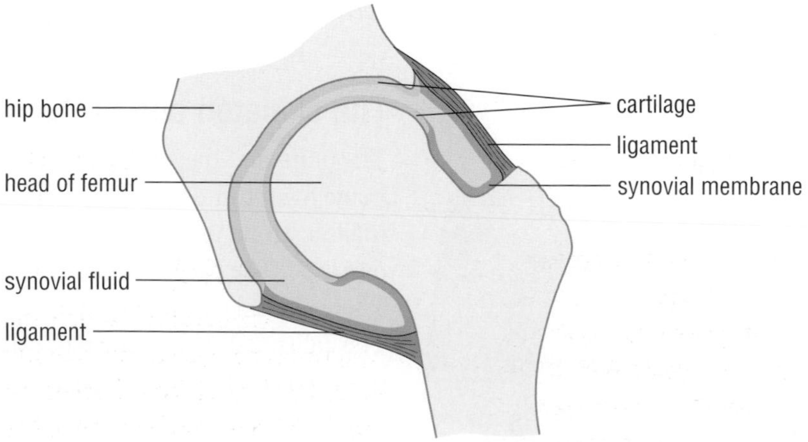

---

hip bone

head of femur

synovial fluid

ligament

cartilage

ligament

synovial membrane

Figure 10.7 Inside a hip joint.

28 Why do you think that some joints are painful in elderly people?

29 How does the body stop you using a damaged joint so that it has time to heal?

To stop the bones coming apart when they move, they are held together by tough fibres called ligaments. To stop them wearing out as they rub over each other, the parts of the bones in the joint are covered with cartilage. This substance has a hard, slippery surface that reduces friction and allows the bones to move over each other easily. In some joints, where there is a lot of movement, cells in a tissue called the synovial membrane make a liquid called synovial fluid. This fluid spreads out over the surfaces of the cartilage in the joint and acts like an oil, reducing friction and wear.

Muscles

Muscle is made up from tissue that has the power to move. It can contract to become shorter. A muscle is attached to two bones across a joint. When a muscle gets shorter it exerts a pulling force. This moves one of the bones but the other stays stationary. For example, in the upper arm the biceps muscle is attached to the shoulder blade and to the radius bone in the forearm. When the biceps shortens or contracts, it exerts a pulling force on the radius and raises the forearm.

A muscle cannot lengthen or extend itself. It needs a pulling force to stretch it again. This force is provided by another muscle. The two muscles are arranged so that when one contracts it pulls on the other muscle, which relaxes and lengthens. For example, in the upper arm the triceps muscle is attached to the shoulder blade, humerus and ulna. When it contracts, the biceps relaxes and the force exerted by the triceps lengthens the biceps and pulls the forearm down. When the biceps contracts again, the triceps relaxes and the force exerted by the biceps lengthens the triceps again and raises the forearm. The action of one muscle produces an opposite effect to the other muscle and causes movement in the opposite direction. The two muscles are therefore called an antagonistic muscle pair.

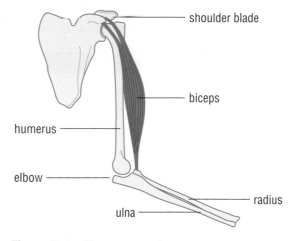

Figure 10.8   Biceps on arm bones.

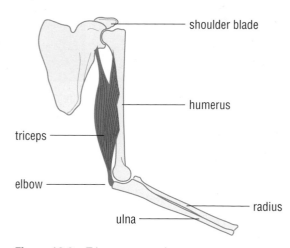

Figure 10.9   Triceps on arm bones.

30 Draw a diagram showing both the biceps and the triceps, with the triceps fully shortened.

31 Using dotted lines, draw the position of the forearm when the biceps is fully shortened.

Drug abuse

Drugs are chemical substances that change the way we think, feel or behave. Some, such as caffeine which is found in tea and coffee, are made by plants. Medical drugs are usually made by the pharmaceutical industry from raw chemical materials. Most drugs are produced to ease the symptoms of a disease or to cure the disease. Although many people think of harmful drugs as being substances like heroin or cannabis, nicotine in cigarettes and alcohol in drinks like beer and wine are also drugs that can harm the body.

People begin taking harmful drugs for a variety of reasons. Some begin because they are unhappy, lonely or feel that they are unable to cope with life. They think the drugs will make them feel better. Others take them because their friends are trying them and they find it difficult to say no (peer pressure). Other people take them because they think it is exciting to use substances that are illegal. Whatever the reason, taking harmful drugs can be dangerous.

WHEN MY MATE OFFERED ME DRUGS, FRANK TOLD ME WHAT TO DO. FRANK

0800 77 66 00 talktofrank.com

FOR INFORMATION, HELP AND ADVICE ABOUT DRUGS TALK TO FRANK

Figure 10.10 FRANK is a Government campaign that offers advice on how drugs can affect you.

General body changes in drug abuse

A drug has an effect on how the body works. As a person continues to take certain drugs the body becomes more tolerant of them and larger amounts of the drug have to be taken for the person to feel its effects. The drug-taker's brain or body generally gets so used to the drug that it becomes changed in some way and becomes physically dependent on the drug. This is known as an addiction. If the person stops taking the drug the body reacts in a range of painful ways, including sickness. These reactions are called withdrawal symptoms.

While the body is becoming physically dependent on the drug, the person may be becoming psychologically dependent on it. This means that they become upset if they are not taking it and develop the irrational fear that they cannot cope with life without the drug.

Dangers of drugs

Ecstasy

Ecstasy affects the body's co-ordination. It can make a person confused. It also acts as a stimulant and can seriously affect people who suffer from epilepsy or have a heart condition. It can be fatal.

Amphetamine or speed

Amphetamines are stimulants. Their use can lead to mental disorders and heart damage.

Cocaine and crack

Cocaine is sniffed or injected. Crack is a form of cocaine that can be smoked. Both are stimulants and are highly addictive. They can make users feel sick, itchy, suffer nose damage, have difficulty sleeping and develop mental disorders.

LSD (acid)

LSD affects the brain and makes the user see things that are not there. These illusions are called hallucinations. They may be frightening and lead to the person being upset when the effect has worn off. People who use LSD regularly can become less alert to the world around them.

Cannabis

Cannabis can produce hallucinations and make people upset. If it is smoked, it can cause the same diseases as smoking tobacco.

Heroin

Heroin can make first-time users so sick that they will never try it again. Regular heroin users are physically and psychologically dependent on it and will commit crimes to obtain money to buy more. They may have slurred speech and seem slightly sleepy. Heroin addicts who inject the drug are in danger of catching hepatitis and AIDS from sharing dirty needles. When heroin addicts try to give up the drug they have great difficulty losing their psychological dependence. They may still crave the drug after they have stopped taking it for a long time.

> ## For discussion
>
> What might be the effect of making the drugs in this section legal?
>
> How effective are posters at preventing drug abuse?
>
> What else could be useful?

Summary

- Regular exercise helps the body to stay healthy (*see page 92*).
- Smoking damages the respiratory system (*see page 93*).
- You can plan a healthy diet if you think of the different foods arranged in a pyramid (*see page 94*).
- Alcohol abuse produces changes in behaviour which can be life-threatening. Prolonged abuse causes liver damage (*see page 95*).
- Solvents damage the lungs, nervous system, kidneys and bone marrow. They cause changes in behaviour that can be life-threatening (*see page 95*).
- Diet can affect the supply of blood to the heart (*see page 96*).
- Muscles and bones work together to provide movement (*see pages 97–99*).
- Illegal drugs can cause damage to a wide range of body organs and can lead to mental disorders (*see pages 99–100*).

End of chapter questions

1 How is someone putting their health at risk when they smoke tobacco daily?

For discussion

What is an ideal lifestyle for a healthy life?

2 Make a table about the drugs featured on page 101 under the headings
- Drugs
- Effects on the body
- Effects on behaviour

3 Design a poster based on the information in your table. Compare the posters produced in your class, and rank them for effectiveness of putting people off trying drugs.

4 Some people have to inject themselves once a day to stay healthy. Can you think of a condition where this is the case, and name the substance that is injected?

How green plants live

How experiments build up information

Scientific processes are not usually understood by a single activity or even one repeated several times for checking. Processes are worked out over many years by a large number of different experiments that require different apparatus and different techniques. They may include making observations, thinking up new ideas and making models to test ideas. The activities form part of a line of research that may go back many years. The results of each experiment may contribute something to our understanding of how a process works. Eventually the results of a large number of different activities may show how the process works. On the following pages, a series of experiments are presented as very simple examples of how their results contributed to our understanding of how plants make food.

If you become a scientist you will use some of the features you read about here, such as looking at the work of others or learning a technique to use in your experiments, in addition to making investigations following the scientific method.

The willow tree experiment

In the 17th century Joannes Baptista van Helmont (1580–1644) performed an experiment on a willow tree. He was interested in what made it grow. At that time scientists believed that everything was made from four 'elements': air, water, fire and earth. Van Helmont believed that water was the most basic 'element' in the universe and that everything was made from it. He set up his experiment by weighing a willow sapling and the soil it was to grow in. Then he planted the sapling in the soil and provided it with nothing but water for the next five years. At the end of his experiment he found that the tree had increased in mass by 73 kilograms but the soil had decreased in mass by only about 60 grams. He concluded that the increase in mass was due to the water the plant had received.

If we were to summarise van Helmont's conclusion it could look like this:

water → mass of plant

Revising the work so far

As plants are food for animals, the simple equation could be rewritten as:

water → food in the plant

Moving on

The idea of food in the plant could then be investigated. A reasonable place to start could be with a plant part that is used as food – the potato.

1 How fair was van Helmont's experiment? Explain your answer.
2 Did the result of the experiment support van Helmont's beliefs? Explain your answer.
3 If you were to repeat van Helmont's experiment, how would you improve it and what table would you construct for recording your results?

Examining a potato with a microscope

If a small slice of potato is examined under the microscope, the cells are found to contain colourless grains. When dilute iodine solution is added to the potato slice the grains turn blue–black. This test shows that the grains are made of starch.

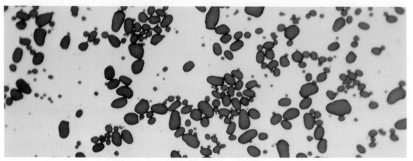

Figure 11.1   Starch grains viewed under a microscope.

Moving on

Having established that plant tubers such as the potato contain starch, it may then be reasonable to try and find out if other parts of the plant contain starch. As the leaves are a major feature of most plants, the search for starch in leaves would be the next task.

Testing a leaf for starch

Iodine does not produce a colour change when it is placed on a leaf because the cell walls will not allow the iodine into the cells and the green pigment masks any colour change. However, the work of others has shown that a leaf can be tested for starch if it is first treated with boiling water and ethanol. The boiling water makes it easier for liquids to leave and enter the plant cells and the ethanol removes the green pigment, chlorophyll, from the leaf and makes the leaf crisp.

If the leaves of a geranium that has been growing on a windowsill or in a greenhouse are tested, they will be found to contain starch.

Revising the work so far

Starch belongs to a nutrient group called carbohydrates, so perhaps the simple equation could be altered to:

water → carbohydrate (starch) in the plant

Moving on

Stephen Hales (1677–1761) discovered that 'a portion of air' helped a plant to survive and Jan Ingenhousz (1730–1799) showed that green plants take up carbon dioxide from the air when they are put in the light. It was also known that water contains only the elements hydrogen and oxygen while carbohydrates contain carbon, hydrogen and oxygen. All this information led to a review of van Helmont's idea that only water was needed to produce the carbohydrate. The review began by considering what else was around the plant apart from water. It was known from van Helmont's work that the soil contributed only a very small amount to the increased mass of the plant. The only other

4 What was the purpose of putting the leaf
  a) in boiling water,
  b) in ethanol?

material coming into contact with the plant was the air. Ingenhousz's work suggested that the carbon dioxide in the air was important. This idea can be tested in the laboratory.

Destarching a plant

If you want to see whether starch has been made, you have to start with a plant that does not have starch. If a plant that has leaves containing starch is left in darkness for two or three days, then tested again, it will be found that the leaves are starch-free. The plant is described as a destarched plant. It can be used to test for the effect of carbon dioxide.

Investigating the effect of carbon dioxide on starch production

Soda lime is a substance that absorbs carbon dioxide and takes it out of the air. Sodium hydrogencarbonate solution is a liquid that releases carbon dioxide into the air.

5 What does soda lime do to the air inside the plastic bag?
6 What does sodium hydrogencarbonate do to the air inside the plastic bag?

Figure 11.2 Plants set up to investigate the effect of carbon dioxide on starch production.

Two destarched plants were set up under transparent plastic bags that were sealed with an elastic band. Before covering the plants with the bags, a small dish of soda lime was added to one plant and a small dish of sodium hydrogencarbonate solution was added to the other. Both plants were left in daylight for a few hours before a leaf from each of them was tested for starch.

The leaf from the plant with the soda lime dish did not contain starch but the leaf from the plant with the sodium hydrogencarbonate did contain starch. This suggested that carbon dioxide is needed for starch production.

Revising the work so far

After reviewing the result of the effect of carbon dioxide on starch production, the simple equation can be modified again to:

carbon dioxide + water → carbohydrate (starch) in a plant

Moving on

Joseph Priestley (1733–1804) studied how things burn. At that time scientists used the phlogiston theory to explain how things burned. They believed that when materials such as wood burned they lost a substance called phlogiston. When a candle burned in a closed volume of air, such as the air in a bell jar, they believed that the candle eventually went out because the air had become filled with phlogiston. It had become phlogisticated.

When Priestley put a plant in the air in which a candle had burned he found that later on a candle would burn in it again. He reasoned that the plant had taken the phlogiston out of the air and had made dephlogisticated air. Later, Ingenhousz re-examined Priestley's results and found that the phlogiston theory was wrong. The plants had in fact produced oxygen.

Water plants can be used to investigate the gases produced by plants because the gases escape from their surface in bubbles that can be easily seen and collected.

Investigating oxygen production in plants

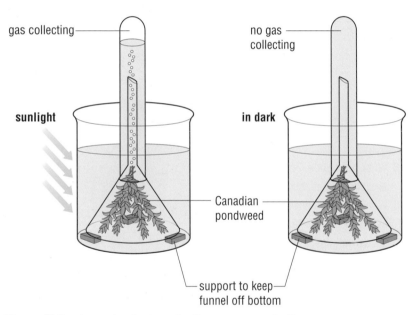

Figure 11.3 Apparatus for investigating oxygen production.

Two samples of Canadian pondweed were set up as shown in Figure 11.3. One was put in a sunny place and the other was kept in the dark. After about a week the amount of gas collected in each test-tube was examined. The plants in the dark had not produced any gas. The plants in the light had produced gas and when it was tested with a glowing splint the splint relighted, showing that the gas contained more oxygen than normal air.

Revising the work so far

From the result of this experiment the equation can be further modified to:

carbon dioxide + water → carbohydrate + oxygen

Moving on

Having established that the carbohydrate starch is formed in leaves it may seem reasonable to find out what affects the presence of starch in leaves. What is it in the plant that allows the reaction to happen? Does the reaction happen all the time or only at certain times of the day or night?

Testing the effect of light on a destarched plant

Two leaves of a destarched plant were set up as shown in Figure 11.4 and left for over four hours in daylight. After that time they were removed and tested for the presence of starch. The leaf kept in the clear plastic sheet contained starch. The leaf kept in the aluminium sheet did not contain any starch. This suggested that light is needed for starch to form in a leaf.

Revising the work so far

After reviewing the result of the effect of light on the leaf, the equation can be modified to:

$$\text{carbon dioxide} + \text{water} \xrightarrow{\text{light}} \text{carbohydrate} + \text{oxygen}$$

Light provides the energy for the chemical reaction to take place. Some of the energy is stored as chemical energy in the carbohydrate.

Moving on

Having discovered a connection between the leaf, light and starch production it may seem reasonable to find out which part of the leaf is important. As most leaves are green it may be suggested that the green pigment, chlorophyll, which is found in chloroplasts of the leaf, is important. If it is lacking, starch should not be made. This hypothesis can be tested by using a variegated leaf, which has some cells that do not have chlorophyll thus making parts of the leaf appear white.

Investigating chlorophyll and starch production

A destarched variegated plant was left in daylight for over four hours. A leaf was then removed and tested for starch. The parts that were green contained starch but the parts that were white did not contain starch. This suggested that chlorophyll is needed for the leaf to produce starch.

Figure 11.4 A destarched plant with two leaves covered by clear plastic and aluminium.

— destarched leaf
— clear plastic covering
— aluminium covering

Figure 11.5 A variegated pelargonium called Lady Plymouth.

After reviewing the result of the effect of chlorophyll on starch production, the equation can be modified to:

$$\text{carbon dioxide} + \text{water} \xrightarrow[\text{chlorophyll}]{\text{light}} \text{carbohydrate} + \text{oxygen}$$

Further experiments showed that the carbohydrate starch was built up in stages from subunits of a substance called glucose. The equation for starch production, or photosynthesis, is now written as:

$$\text{carbon dioxide} + \text{water} \xrightarrow[\text{chlorophyll}]{\text{light}} \text{glucose} + \text{oxygen}$$

The fate of glucose in plants

Glucose made in the process of photosynthesis is used to release energy in the process of respiration. Glucose is also used to make many other molecules in the plant. It may be used to make cellulose for the cell walls or turned into fats for the cell membranes or for storage in seeds. When glucose is made in large quantities in the leaf cells it is converted into starch and stored. When the level of glucose in the cells becomes low the starch is converted back into glucose.

Glucose is also used to make the sugars in fruits so that their sweet taste makes them attractive to animals. Nitrogen and other minerals join with the elements in glucose to make amino acids and proteins.

Figure 11.6 summarises all these uses of glucose.

7 Try to describe photosynthesis in your own words.

8 When will the amount of glucose in a leaf cell rise to a high level? Explain your answer.
9 Why does starch form?
10 When will the amount of starch in a leaf cell decrease? Explain why this happens.

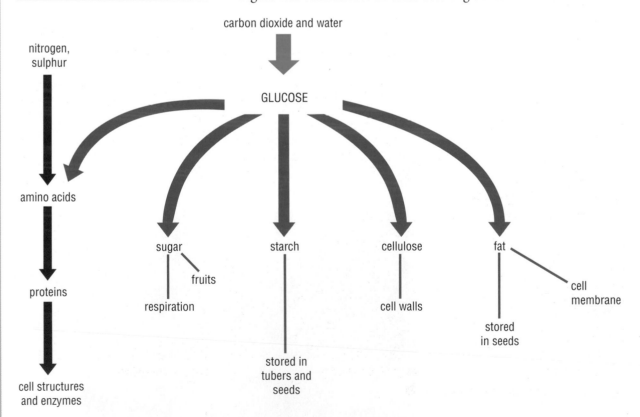

Figure 11.6 Schematic drawing of the fate of glucose.

11 What do plants take from the air and give to the air when they respire?
12 What do plants take from the air and give to the air during photosynthesis?
13 Compare the equations for photosynthesis and respiration.
14 How does the amount of
  a) carbon dioxide and
  b) oxygen vary around a plant over a 24-hour period? Explain your answer.

Plant respiration

Plant cells need energy to drive their life processes. As in animals, this energy is released in respiration:

$$\text{glucose} + \text{oxygen} \rightarrow \text{carbon dioxide} + \text{water} + \text{energy}$$

Plants respire 24 hours a day. They take in oxygen and produce carbon dioxide. During daylight, photosynthesis also takes place. In this reaction carbon dioxide is used up and oxygen is produced. In bright sunlight the speed at which plants produce oxygen is greater than the speed at which they use up oxygen in respiration.

Use of mineral salts by plants

When chemists began studying plants they discovered that they contained a wide range of elements. With the exceptions of carbon, hydrogen and oxygen, the plants obtained these elements from mineral salts in the soil.

The importance of each element to the health of plants was assessed by setting up experiments in which the plants received all the necessary mineral salts except for the one under investigation. From these studies it was found that:

15 Put the information about mineral salts and their uses by plants into a table. Include information about what happens if the mineral salt is missing.
16 What minerals might be missing if the leaves go yellow?
17 Why might a plant show poor growth?

- nitrogen is taken in as nitrates and is needed to form proteins and chlorophyll. Without nitrogen the plant's leaves turn yellow and the plant shows poor growth.
- phosphorus is needed to help the transfer of energy in photosynthesis and respiration. Without phosphorus a plant shows poor growth.
- potassium helps the plant to make protein and chlorophyll. If it is lacking the leaves become yellow and grow abnormally.

Lack of nitrogen

Lack of phosphorus

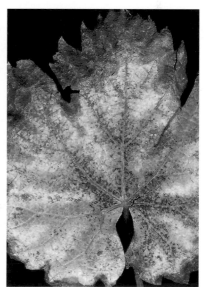
Lack of potassium

Figure 11.7 Plants showing mineral deficiency.

Path of minerals through living things

When animals eat plants, they take in the minerals and use them in their bodies. Some of the minerals are released in the solid and liquid wastes that animals produce. As bacteria feed on these wastes, the mineral salts are released back into the soil. The mineral salts are also released when the plants and animals die and microbes break down their bodies in the process of decomposition. Plants, animals and their wastes are biodegradable. This means they can be broken down into simple substances that can be used again to make new living organisms. These simple substances have been recycled since the beginning of life on Earth.

Water and minerals in a plant

Most plant roots have projections called root hairs. The tips of the root hairs grow out into the spaces between the soil particles. There may be up to 500 root hairs in a square centimetre of root surface. They greatly increase the surface area of the root so that large quantities of water can pass through them into the plant. The water in the soil is drawn into the plant to replace the water that is lost through evaporation from the leaves. The plant does not have to use energy to take the water in.

Mineral salts are dissolved in the soil water. The plant has to use energy to take them in. This energy is provided by the root cells when they use oxygen in respiration. The roots get the oxygen from the air spaces between the soil particles.

> **18** Could you be a recycled dinosaur? Explain your answer.

> **19** If a plant is over-watered all the spaces between the soil particles become filled with water. How does this water-logged soil affect
> **a)** the plant roots,
> **b)** the plant's growth?

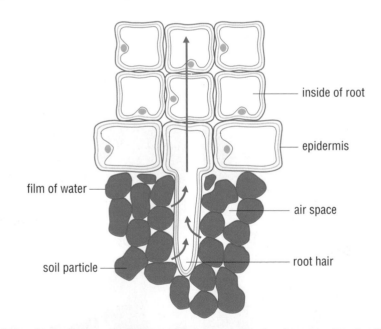

Figure 11.8 Schematic drawing of the movement of water and mineral salts in the root of a plant.

inside of root

epidermis

film of water

air space

soil particle

root hair

Carnivorous plants

Some plants live in conditions where minerals, particularly nitrogen, are unavailable. They are therefore unable to take up these minerals from the soil. The leaves of these plants have adapted to allow them to trap and kill animals, before digesting their bodies to obtain nitrogen from the animal protein.

Butterwort, sundew and Venus fly trap grow in peaty bogs. Bladderwort is found in ponds.

The butterwort has leaves arranged in a rosette around the flower stalks. Each leaf is from 2 to 8 cm long. On the upper surface of the leaf are hairs that secrete a liquid containing protein-digesting enzymes. The edges of the leaf turn up to make a lip that prevents the sticky liquid from flowing away into the soil. If an insect lands on the liquid it cannot escape. Its soft parts are digested and absorbed into the leaf.

The leaves of the sundew have long stalks and circular blades. The upper surface of the leaf blade is covered in hairs that secrete sticky drops of a liquid that contains protein-digesting enzymes. When an

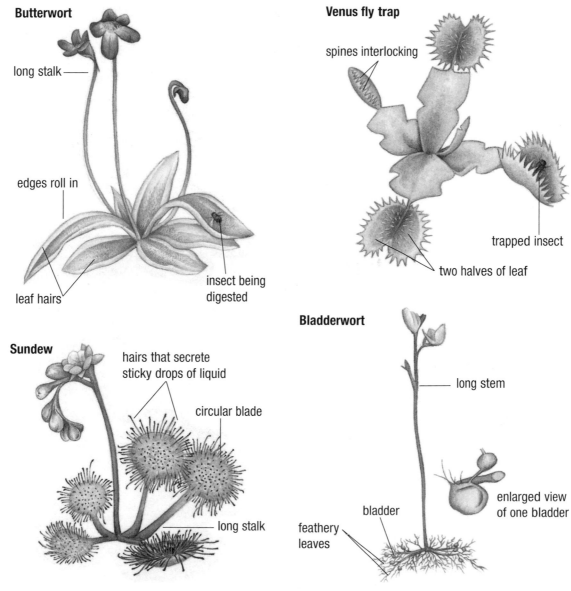

Figure 11.9 Carnivorous plants.

insect lands on the leaf it sticks to the hairs and the enzymes digest its soft parts, leaving the hard parts to be blown away by the wind.

Each leaf of the Venus fly trap is divided into two halves that can spring together in 0.03 seconds. There are three hairs on each half of the leaf which act as triggers. If an insect lands on the leaf and touches them the trap is sprung. The spines on the edges of the leaf interlock and stop the insect escaping. A liquid is secreted from the leaf's surface which digests and absorbs the insect's body. In 24 hours the leaf opens again. After it has digested four insects the leaf dies and is replaced by a new one.

The bladderwort is an aquatic plant that has pin-head sized traps on its feathery leaves. Each trap is called a bladder. The bladder is a globe-shaped structure that has a cavity which can be filled with water or air. The trap is set by removing the water from the cavity. Inside each bladder are cells that absorb water. At the bladder's mouth is a trap door, trigger hairs and cells that contain sugar to attract prey. When a water flea touches the trigger hairs, the trap door opens and water rushes into the bladder carrying the water flea with it. The door closes behind it and the water flea is absorbed within a few days.

Oxygen and carbon dioxide in the atmosphere

About 20% of the atmosphere is composed of oxygen and about 0.03% is composed of carbon dioxide. These two amounts remain almost the same from year to year. The reason they do not change much is that the carbon dioxide produced by animals and plants in respiration is used up in photosynthesis, and the oxygen produced by the plants is used up by plants and animals in respiration. However, a small annual increase in carbon dioxide in the atmosphere has occurred in modern times due to the burning of fossil fuels, and this is believed to be a major cause of global warming (see page 205).

Figure 11.10 Some of the oxygen that these deer are breathing has been produced by the trees around them.

The carbon cycle

The carbon dioxide taken into a plant is used to make glucose, which may be transported to a storage organ and converted into starch. If the storage organ is a potato, for example, it may be dug up out of the ground, cooked and eaten. The starch is broken down in digestion to glucose and taken into the blood. In the body the glucose may be used for respiration and the carbon is released as carbon dioxide. If too much

22 How successful are the plant traps? Explain your answer.
23 What methods are used by carnivorous plants to break down an insect's body?

24 Why do humans not suffocate at night when the plants around them cannot photosynthesise?
25 What effect will reducing the number of plants on the surface of our planet have on the animals?

26 Read the account of the carbon cycle and draw the paths that carbon can take.
27 Now add in the path that the carbon would take if the potato plant died.
28 Why is the path called the carbon cycle?

high-energy food is being eaten the glucose may be converted into fat and the carbon remains in the body. When the body dies, microbes feed on it and break it down into simple substances. The microbes thus release the carbon back into the air, as carbon dioxide, when they respire.

Summary

- Iodine solution is used in the test for starch (*see page 104*).
- Boiling water and ethanol are used, with care, to remove chlorophyll from a leaf (*see page 104*).
- A plant is destarched by leaving it in the dark for 2 or 3 days (*see page 105*).
- Water and carbon dioxide are the raw materials of photosynthesis (*see page 105*).
- Light and chlorophyll are needed for a plant to photosynthesise (*see page 108*).
- Carbohydrate (glucose and starch) and oxygen are the products of photosynthesis (*see page 108*).
- Glucose made in the process of photosynthesis is used to release energy by respiration (*see page 109*).
- Mineral salts are needed for healthy plant growth (*see page 109*).
- Water and minerals enter the plant through the root hairs (*see page 110*).
- Photosynthesis and respiration keep the levels of oxygen and carbon dioxide in the air fairly constant (*see page 112*).
- Carbon passes from the air to a plant, then to an animal and finally a microbe releases it into the air again as it moves around the carbon cycle (*see page 112*).

End of chapter questions

How does a growing cucumber's weight change in a day? Large cucumbers appear to grow quickly. In this experiment a cucumber plant was placed close to a top-pan balance and one of its growing cucumbers was placed on the pan. The weight of the cucumber was measured every hour between 9.00 am and 4.00 pm. The weight was displayed in the graph in Figure 11.11.

1 What was the gain in weight over the seven-hour period?

2 Construct a table to display the increase in weight in each of the seven hours.

3 When was the period of
   a) greatest and
   b) least growth?

4 If the cucumber plant had some of its leaves removed before the experiment, how would you expect its growth graph to compare with the graph in this experiment? Explain your answer.

> **For discussion**
>
> The rainforests have been described as the world's lungs. What do you think this means?

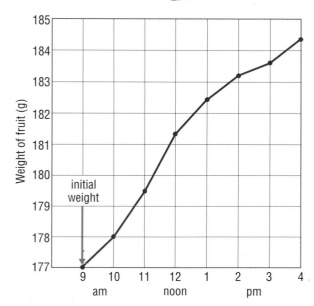

Figure 11.11

HOW GREEN PLANTS LIVE

113

Plants for food

You may start your day with a breakfast like this one.

corn flakes

golden, crisp, light
flakes of corn

500 gram ℮

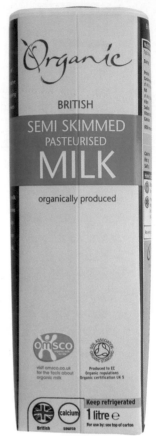

Figure 12.1 Examples of some breakfast foods.

1 Write a food chain for each item in this meal.
2 Write down the items in your breakfast. Make a food chain from each of these items. How many food chains feature animals as well as plants?
3 Read through Chapter 5 again and assess the healthiness of your breakfast. If it is unhealthy what could you do to change it?

Where do all these foods come from? The cereal flakes are made from grains of maize – the grains you also eat as corn on the cob. The milk comes from a cow, but what does the cow eat? The slice of bread is made from the flour of another cereal called wheat. It is spread with a margarine made from sunflower oil, and honey that is made from the nectar of flowers by bees.

What parts of a plant are eaten?
Seeds and fruits

If you look at Figure 1.17 on page 12 you can see that an ovule is surrounded by an ovary wall. After fertilisation takes place, the ovule forms a seed while the ovary wall becomes the fruit, both of which are often eaten.

Cereals

Cereals are crops that have been selectively bred from species of grass. A cereal grain is a fruit containing a single seed. The fruit forms a thin, dry coat around the seed. In wheat, the fruit is called the bran. The fruit is usually removed from the grain before the grain is used to make food.

The main cereal crops grown in temperate climates are:

- wheat – ground up to make flour for a wide range of food products
- maize – for breakfast cereals, sweet corn, corn on the cob
- barley – used in soups, stews and for farm animal food
- oats – used for making porridge and for farm animal food
- rye – used to make certain types of bread and biscuits.

The main cereal crops grown in tropical climates are:

- rice – simply boiled or fried, or made into rice pudding
- millet – mixed with water to make a type of porridge.

Legumes

The main crops that produce seeds for food are the legumes. Peas, beans and lentils are seeds and are sometimes called pulses. They are produced in a fruit called a pod. In some kinds of legumes, such as runner beans and mangetout, the pod and the seeds are eaten. The peanut is an unusual legume. After fertilisation the fruit and seeds develop under the soil. The fruit forms a woody shell around the seeds that we call peanuts.

Nuts

A true nut is made from a fruit which has developed into a woody wall around the single seed. The brazil nut is an example of a nut with a very strong woody wall which must be cracked to release the seed – the part we call the brazil nut.

Fruits

In many plants after fertilisation the ovary wall swells up and becomes fleshy or succulent. Its surface may become a bright colour to make the fruit look attractive enough to eat. Some fruits such as the plum or cherry contain just one seed which has a tough woody coat. This seed is known as the stone. Other fruits such as the orange and the tomato have a centre which is full of small seeds. The apple and the pear are not really fruits and are known as false fruits. The fleshy part which makes the fruit does not form from the ovary wall, but forms from the top of the flower stalk which swells up and grows round the ovary and its seeds.

Vegetables

In everyday life, beans, lentils, peas and tomatoes are all called vegetables. There are many other plants in this group of foods. Each vegetable plant is grown so that a particular part of its body can supply nourishment. Table 12.1 shows the parts of some vegetables that are used as food.

4 Use Table 5.2 on page 46 to answer this question.
   a) Compare the nourishment provided by peas and lentils.
   b) Could either be used as a substitute for chicken? Explain your answer.

5 Which parts of the following plants are used as food – radish, leek, parsnip, cress, cucumber, chicory, kohlrabi?

Table 12.1

Vegetable	Part of plant
broccoli, cauliflower	flower
asparagus	stem
Brussels sprout	bud
celery	leaf stalk
lettuce, onion	leaves
potato	tuber
carrot, beetroot, cassava, yam	root

Fertilisers

When plants are grown for food, they are eventually harvested and taken out of the soil. This means that the minerals they have taken from the soil go with them to market, and there are therefore fewer minerals left in the soil for the next crop. Fertilisers are added to the soil to replace the minerals that have been taken away in the crop. They are also added in quantities that will make sure that the plants grow as healthily as possible and produce a large crop. The amount of fertiliser added to the soil has to be carefully calculated. If too little is added, the amount of food produced by the crop (called the yield) will be small. If too much is added, the plants will not use all the minerals and they may be washed into streams and rivers and cause pollution (see page 80). There are two kinds of fertiliser. Inorganic fertilisers such as ammonium nitrate are manufactured chemical compounds. Organic fertilisers are made from the wastes of farm animals (manure) and humans (sewage sludge).

Inorganic fertilisers can give crops an almost instant supply of minerals, as the minerals dissolve in the soil water as soon as they reach it, and can be taken up by the roots straight away. The minerals in manure are released more slowly as decomposers (see page 77) in the soil break it down.

Inorganic fertilisers are light in weight and so can be spread from aeroplanes and helicopters flying over the crops. This means that they can be applied to the crop at any time without damaging the crop. Manure is too heavy to be spread in this way and must be spread from the trailer attached to the tractor. If the manure was spread while the crop was growing, the tractor and trailer would damage the crop, so the manure must instead be spread on the soil before the crop is sown. This also allows some time for the manure to release minerals into the soil.

Fertilisers and soil structure

When manure is added to soil, it adds humus and this helps to bind the rock particles together and keep the soil crumbs large. If the soil receives only inorganic fertilisers the humus is gradually lost and the soil crumbs break down. The rocky fragments that remain form a dust which can be easily blown away by the wind.

Pests and pesticides

Fertilisers are used to make the crop yield as high as possible. The crop plants may however be affected by other organisms, which either compete with them for resources, or feed on them. These organisims are known as pests, and chemicals called pesticides have been developed to kill them. There are three kinds of pesticides – herbicides, fungicides and insecticides.

The problem with weeds

When a crop is sown, the seeds are planted so that the plants will grow a certain distance apart from each other. This distance allows each plant to receive all the sunlight, water and minerals that it needs to grow healthily and produce a large yield.

A weed is a plant growing in the wrong place. For example, poppies may be grown in a flower bed to make a garden look attractive, but if

6 What kind of fertiliser gives you the greater control over providing minerals for a crop – inorganic or organic? Explain your answer.

7 If a farmer uses only inorganic fertilisers how may the soil organisms be affected?

Figure 12.2 Poppies growing as weeds in a field of wheat.

they grow in a field of wheat they are weeds because they should not be growing there. Weeds grow in the spaces between the crop plants and compete with them for sunlight, water and minerals. This means that the crop plants may receive less sunlight because the weeds shade them, and receive less water and minerals because the weed plants take in some for their own growth. Weeds can also be infested with microbes which can cause disease in the crop plant. For example, cereals may be attacked by fungi that live on grass plants growing as weeds in the crop.

Herbicides

Weeds are killed by herbicides. There are two kinds of herbicide – non-selective and selective. A non-selective herbicide kills any plant. It can be used to clear areas of all plant life so that crops can be grown in the soil later. It must not be used when a crop is growing as it will kill the crop plants too. A selective herbicide kills only certain plants – the weeds – and leaves the crop plants unharmed. It can be used when the crop is growing.

Herbicides may be sprayed onto crops from the air. Some of the herbicide may drift away from the field and into surrounding natural habitats. When this happens, the herbicide can kill plants there. Many wild flowers have been destroyed in this way.

> **8** How does the use of herbicides on a farm affect the honey production of a local bee keeper? Explain your answer.

Fungicides

A fungicide is a substance that kills fungi. Fungal spores may be in the soil of a crop field or floating in the air above it. They may even settle on the seeds before they are sown. Fungicides are coated on seeds to protect them when they germinate. They are also applied to the soil to prevent fungi attacking the roots, and are sprayed on crop plants to give them a protective coat against fungal spores in the air.

Insecticides

When a large number of plants of the same kind are grown together they can provide a huge feeding area for insects. Large populations of insects can build up on the plants and cause great damage. Insecticides are used to kill these insect pests. If insecticides drift away from the fields after spraying, they can kill other insects in their natural habitats. They can also move along the food chain.

> **9** You have a field of weeds and need to plant a crop in it. How could you use pesticides to prepare the ground for your crop and protect it while it is growing?

A poison in the food chain

In 1935 Paul Müller (1899–1965) set up a research programme to find a substance that would kill insects but would not harm other animals. Insects were his target because some species are plant pests and devastate farm crops and others carry microbes that cause disease in

The greenhouse

For a large part of the year the plants in a greenhouse receive all the light they need from the Sun. The growing season can be extended by providing extra light from special lamps. Plants use certain wavelengths of sunlight for photosynthesis and these lamps emit light at these wavelengths.

Figure 12.6 A greenhouse with lamps and ventilators.

For discussion

It has been proposed to construct a huge greenhouse at a farm close to a village. How do you think the villagers will receive the news?

12 How do you think the action of
  a) the lamps and
  b) the ventilators are controlled?

In summer, a greenhouse may receive all the heat it needs from the Sun too. In winter, heaters can be used to lengthen the growing season. The amount of carbon dioxide in the air in a greenhouse can be increased by pumping in extra carbon dioxide. In winter, paraffin heaters can be used to supply extra carbon dioxide to the air as they warm it up.

While a crop is growing the conditions need to be kept as constant as possible. Very high light intensities from the Sun on a clear summer day can damage chlorophyll, and this in turn reduces the rate of photosynthesis. This can be prevented by having blinds which extend automatically when the light intensity increases to a damaging level. If the greenhouse becomes too hot ventilators can open automatically to allow the hot air to escape and cooler air to enter.

Summary

- Plants provide us with food in the form of seeds, cereal grains, nuts, fruits and vegetables (*see pages 114–115*).
- Fertilisers provide minerals for plant growth (*see page 116*).
- Pesticides remove organisms that can damage crops (*see page 116*).
- Poisons can move through food chains (*see page 117*).
- Plant growth can be controlled by regulating factors such as the amount of light, heat or carbon around a plant (*see pages 118–120*).

End of chapter question

1 a) Write down twelve foods that you frequently eat.
  b) Work out a food chain for each one.
  c) On which food plants does your diet greatly depend?

chemistry

13 *Introducing chemistry*

What is chemistry?

Chemistry is the study of the structure of substances and how they change. It developed out of a human activity called alchemy, which was practised in Europe, China and India for over a thousand years.

Alchemy was the study of matter, but the ideas the alchemists used in their work were not based on scientific investigations. They believed that there was a substance called the philosopher's stone which could change metals such as lead into gold. Alchemists performed experiments on a wide range of substances to try to find the philosopher's stone. They kept notes of their work, but used strange symbols to keep their work secret. The mysterious way in which they worked, and the coloured flames, explosions, smoke and fumes they made, meant they became known as magicians and wizards. None of them ever found the philosopher's stone, and by the 17th century scientific investigations had replaced the alchemists' experiments. Some of the alchemists' knowledge was used by the first chemists, who based their conclusions on what they observed and not on ideas about changing lead into gold.

Today, chemistry and the work of chemists affects our lives in many ways; from the paper, ink and glue in this book, to the food in the last meal you ate and the fibres and colours of the clothes you are wearing now.

> 1 One of the first observations of how things change was made by people watching a fire burn. What changes occur when wood burns?
>
> 2 Why do you think that alchemists wanted to keep their work secret?

Measuring quantities

For many investigations, the quantities of substances taking part in a chemical reaction need to be known and also the quantities of the substances that are produced. The next few pages consider the basic measurements made in chemistry.

Figure 13.1   An alchemist at work.

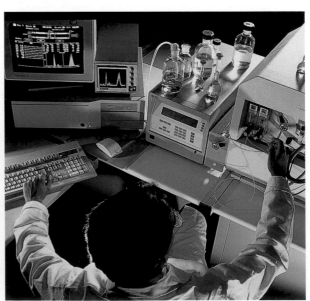

Figure 13.2   A present-day chemist at work.

Measuring volumes of liquids

The volume of a liquid can be found by pouring it into a measuring cylinder and reading the scale. A measuring cylinder can also be used to prepare a specified volume of a liquid by pouring in an amount of the liquid, then either topping it up or pouring some out, until the required volume is present in the cylinder.

The scales on measuring cylinders indicate the volume in either millilitres (ml) or cubic centimetres (cm^3). A millilitre is a thousandth of a litre and is used in measuring out liquids that are sold in bottles and cans. It is also used to indicate the quantities of liquids that are used in recipes. In scientific work the unit cm^3 is used. $1 \ cm^3 = 1 \ ml$.

The surface of the liquid in a measuring cylinder curves upwards at the point where it touches the inside of the cylinder. This curvature is called the meniscus. To read the volume of a liquid accurately, the base of the measuring cylinder must be placed on a flat surface and the eye must be level with the surface of the liquid in the middle of the cylinder (see Figure 13.3). The volume of liquid in the cylinder in Figure 13.3 is 35 cm^3.

The meniscus of mercury is unusual in that it curves downwards. You would never measure mercury in a measuring cylinder (it would be too dangerous), but you can see its meniscus in a mercury-in-glass thermometer.

Figure 13.3 Reading the volume of liquid in a measuring cylinder.

---

3 How would reading the volume of the liquid from the top of the meniscus make the reading inaccurate?

Measuring the volume of a gas

The volume of a gas may be measured using a syringe with a scale marked on it. The scale may measure in millilitres or cubic centimetres. As the gas is produced it passes along a tube into the syringe and pushes out the plunger. The volume collected in the syringe can be measured by reading the scale at the place where the plunger comes to rest (see Figure 13.4).

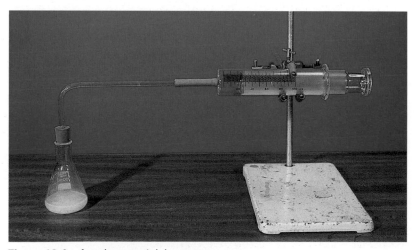

Figure 13.4 A syringe containing a gas.

Measuring the mass of a solid or liquid

The mass of a substance can be found by using a top-loading balance. This piece of equipment is very sensitive and must be treated with great care at all times. Loads for weighing must be put onto, and removed from, the pan carefully.

The mass of the substance in Figure 13.5 is found by reading the main number display and then reading the two decimal places from the second number display. The mass of the substance in Figure 13.5 is 17.24 g.

Figure 13.5 A top-loading balance.

The top-loading balance measures mass in grams (g) but it also has a mechanism which allows larger masses to be measured in kilograms (kg). For most laboratory work the balance is used to measure small masses in grams.

If a substance such as a liquid is to be weighed in a beaker, the mass of the beaker must first be found. The mass of the substance and the beaker is then found, and the mass of the substance is calculated by subtracting the mass of the beaker from this total. For example:

$$
\begin{aligned}
\text{mass of beaker} &= 50.00\text{ g} \\
\text{mass of beaker + substance} &= 120.00\text{ g} \\
\text{mass of substance} &= 120.00 - 50.00 = 70.00\text{ g}
\end{aligned}
$$

Some balances have a tare. This mechanism allows substances to be weighed out without having to make a calculation. It is used in the following way: the empty beaker is placed on the top pan and its mass is displayed. The tare is then used by turning or pushing a control on the side of the balance. This action brings the reading on the balance back to zero, even though the beaker is still on the pan. The substance can then be placed in the beaker and its mass is read directly from the display.

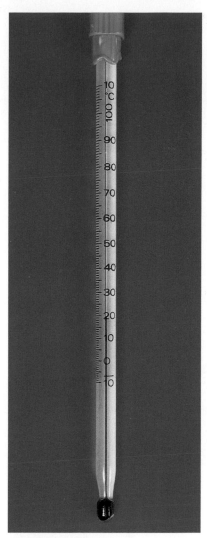

Figure 13.6 A thermometer.

Measuring temperature

The thermometer is used to measure temperature (see Figure 13.6). It is a glass tube, with a small container called a bulb at one end. In the container is a liquid which expands or contracts as the temperature changes. The liquid may be mercury or coloured alcohol.

A small amount of the liquid forms a thread in the thermometer tube. As the temperature rises the liquid expands and the thread in the tube increases in length. When the temperature falls the liquid contracts and the thread decreases in length. The length of the thread is measured on a scale on the tube. The Celsius scale is used on most laboratory thermometers. This scale is used to measure the temperature in degrees Celsius (°C).

The end of the thermometer that does not have the bulb may be capped with a piece of plastic, designed to stop the thermometer rolling along the bench and falling onto the floor. Some people fail to tell the difference between the cap and the bulb when they first use a thermometer and use it the wrong way up!

When the temperature of a liquid is to be measured, the bulb of the thermometer should be put into the liquid and the movement of the mercury or coloured alcohol in the thermometer observed. When the expansion or contraction is finished, the temperature can be read from the scale. While the temperature is being read the bulb of the thermometer must be kept immersed in the liquid. If it is removed the temperature of the air will be recorded.

Apparatus

The equipment that is used in a chemistry laboratory is called apparatus. Many pieces of apparatus are made of glass because it is transparent, so the chemical reactions are easy to see. Glass is also easy to clean. The ordinary glass used in objects found in the home breaks if it is heated. The glass apparatus used in the laboratory is usually made from borosilicate glass, also known as Pyrex. This glass does not break when it is heated. It is also used to make kitchen glassware like casserole dishes, which can be safely put in an oven to cook a meal. Figure 13.7 (overleaf) shows some common pieces of apparatus.

The Bunsen burner

Robert Bunsen (1811–1899) was a German scientist. He made many investigations and his work included the invention of a battery and developing a way of identifying substances from the flames they produced. This method has been developed to identify substances in stars.

Bunsen is best known for the Bunsen burner, although it is uncertain whether, in fact, he invented it. However, he used it so widely in his investigations that other scientists began to use it too. Today it is used in laboratories throughout the world to give a strong steady source of heat without smoke.

By using the burner, Bunsen and a colleague discovered two new elements (see Table 18.1, page 164).

4 What temperature does the thermometer in Figure 13.6 show?

5 Someone is asked to take the temperature of a liquid. They put the bulb of the thermometer in the liquid for a few minutes, then take it out to read it. Will their reading be accurate? Explain your answer.

6 What advice would you give to someone who was taking the temperature of a liquid?

7 What are the advantages of using a burner, compared with a fire or a candle?

8 Who was Bunsen's colleague, and what were the elements they discovered?

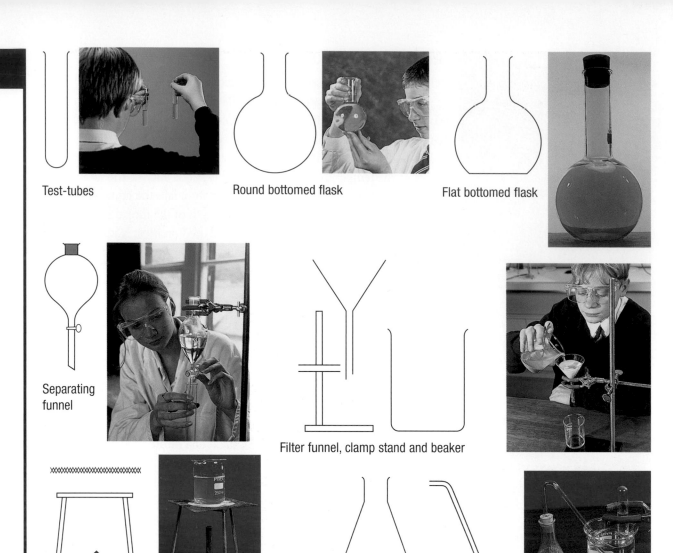

Test-tubes

Round bottomed flask

Flat bottomed flask

Separating funnel

Filter funnel, clamp stand and beaker

heat

Bunsen burner, tripod and gauze

Conical flask with delivery tube

Figure 13.7

Diagrams of apparatus

When a record of an experiment is being made, a diagram of how the apparatus was set up is included. Each piece of apparatus is represented diagrammatically so the way the apparatus was set up can be clearly seen. Figure 13.7 shows the diagrams used to represent common pieces of apparatus.

9 What are the pieces of apparatus represented by the diagrams in Figure 13.8a and b?

a)

b)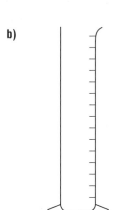

Figure 13.8

10 Figure 13.9 shows how some pieces of apparatus were set up in an experiment. Draw a diagram of them and label each one.

11 What is the volume of liquid in the measuring cylinder in Figure 13.10a?

12 What is the volume of gas in the syringe in Figure 13.10b?

13 What is the mass of the beaker on the balance in Figure 13.10c?

Figure 13.9 Apparatus for a simple experiment.

a)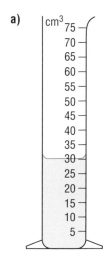

b)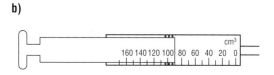

c)

Figure 13.10

More complicated apparatus

Some pieces of apparatus are complicated. For example, the Liebig condenser is used to convert steam into water. Figure 13.11 shows the Liebig condenser set up with other pieces of apparatus to carry out distillation (see pages 159–60). In this apparatus set-up, note how bungs with tubes passing through them are represented diagrammatically.

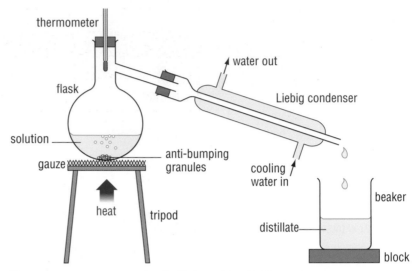

Figure 13.11   Apparatus for distillation with a Liebig condenser.

Warning signs

Like all sciences, chemistry is a practical subject but some of the substances that are used in the laboratory are dangerous if not handled properly. The containers of these substances are labelled with a warning symbol such as those shown in Figure 13.12.

corrosive

explosive

harmful or irritant

highly flammable

oxidising

radioactive

toxic

Figure 13.12   Warning symbols.

14 What do you understand by the words:
   a) corrosive,
   b) irritant,
   c) flammable,
   d) radioactive,
   e) toxic?

Summary

- Chemistry developed from alchemy about 200 years ago (*see page 122*).
- Special apparatus is used to measure the volume, mass and temperature of a substance (*see pages 122, 123 and 124*).
- Laboratory apparatus can be represented in diagrams (*see page 126*).
- Warning signs are used on the containers of dangerous substances (*see page 128*).

End of chapter question

1 What advice can you give to someone to help them to take careful readings in experiments?

The world of matter

Figure 14.1 The three states of matter on a school hike.

The three states of matter are solid, liquid and gas. When you go for a walk you move across the solid surface of the Earth. Your body pushes through a mixture of gases that we call the air. If it rains as you walk along, droplets of liquid fall from the sky. Solids, liquids and gases are the three states of matter on this planet, and for most other places in the Universe too. Not only does your body move through a world made from the three states of matter, it is also made from the three states of matter. Solid bones are moved by solid muscles, while liquids move through your blood vessels and intestines. When you breathe in, air (a mixture of gases) fills your windpipe and lungs.

Properties of matter

You can tell one state of matter from another by examining its properties.

Solids, liquids and gases all have mass and volume. They also have density, which is found by dividing the mass of the substance by its volume. For example, a solid with a mass of 100 g and a volume of

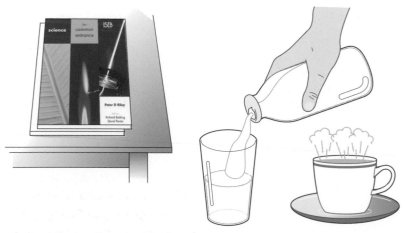

Figure 14.2 A solid, a liquid and a gas.

1 Make a table of the properties of the three states of matter.
2 How are all three states of matter
   a) similar,
   b) different?
3 Calculate the densities of these substances:
   a) a plank of wood used for a shelf that has a volume of 1000 cm³ and a mass of 650 g,
   b) the petrol in a car petrol tank that has a volume of 3000 cm³ and a mass of 2400 g,
   c) the air in a box of 1000 cm³ that has a mass of 1.3 g.

10 cm³ has a density of 100/10 = 10 g/cm³. Another solid with a mass of 200 g and a volume of 10 cm³ has a density of 200/10 = 20 g/cm³. This second solid has a higher density than the first solid.

A solid has got a definite shape and a high density. It is very hard to make it flow or to compress (squash) it. A solid has a definite mass and a volume that does not change. A liquid also has a definite mass and volume. Its density is high and it is hard to compress, but it is easy to make it flow. The shape of the liquid varies and depends on the shape of the container holding it. The shape and volume of a gas vary, and it is easy to make it flow and to compress it. A gas has a definite mass but its density is low.

Using the properties of matter

The different properties of solids, liquids and gases lead to specific uses.

As solids have fixed shapes and volumes and are hard to compress, they are used to build structures that range in size from tiny machines to office tower blocks.

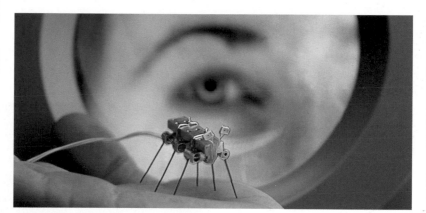

Figure 14.3 A 'robot gnat' developed by nanotechnology.

Liquids flow but are also hard to compress. They are used, for example, as lubricants and as brake fluid (see page 330).

Gases can be compressed; they are used, for example, in aerosol spray cans and to inflate tyres.

Changing states

The state of matter of a substance can be changed. It is changed by heating or cooling.

Melting and freezing

If a solid is heated enough it loses its shape and starts to flow. This change is called melting and the solid turns into a liquid. The temperature at which melting takes place is called the melting point.

If a liquid is cooled enough it loses its ability to flow, forms a shape and turns into a solid. This change is called freezing. The temperature at which freezing takes place is called the freezing point. The temperature of the melting point is the same as the temperature of the freezing point.

Figure 14.4 A tower block under construction in Canary Wharf, London.

Evaporating and boiling

A solid turns into a liquid at one definite temperature but a liquid turns into a gas over a range of temperatures. For example, a drop of water can turn into a gas at room temperature of about 20 °C while outside a puddle of water dries up in the warmth of the Sun. The process by which a liquid changes into a gas over a range of temperatures is called evaporation. The gas escapes from the surface of the liquid. If the temperature of the liquid is raised it evaporates faster. At a certain temperature the gas forms inside the liquid and makes bubbles which rise to the surface and burst into the air. This process is called boiling. The temperature at which it takes place is called the boiling point. If the boiling liquid is heated more strongly its temperature does not rise but it boils more quickly.

Condensation

When a gas is cooled down it turns into a liquid by a process called condensation. This process is the opposite of evaporation. When the water in a kettle boils it forms a colourless gas called steam that rushes out of the kettle spout. A few centimetres above the spout the steam cools and condenses to form a cloud of water droplets which is often wrongly called steam. The real steam cannot be seen and is in the gap between the spout and the base of the cloud of water droplets.

Figure 14.5   A cloud of water droplets from a boiling kettle.

Sublimation

There are a few solids which turn directly into a gas when they are heated. They do not change into a liquid first. This process is called sublimation.

Solid carbon dioxide, known as dry ice, sublimes when it is heated to −78 °C. It can be used on a stage to produce a mist in the air when it warms up (see Figure 14.6).

Figure 14.6   Dry ice being used at a concert.

The term sublimation is also used when a gas turns directly into a solid. Sulphur vapour escaping from a volcano sublimes to form a solid crust on the rocks close by.

Mass and the changes of state

When any substance changes state, such as turning from a liquid to a solid or a liquid to a gas, the mass of the substance does not change.

The changing state of water

It has been estimated that there are 1.5 million million million litres of water on the Earth. Water can change from solid to liquid to gas and back to liquid and solid again at the temperatures found naturally on the Earth. Water moves between the oceans, atmosphere and land in a huge circular path called the water cycle (see Figure 14.7).

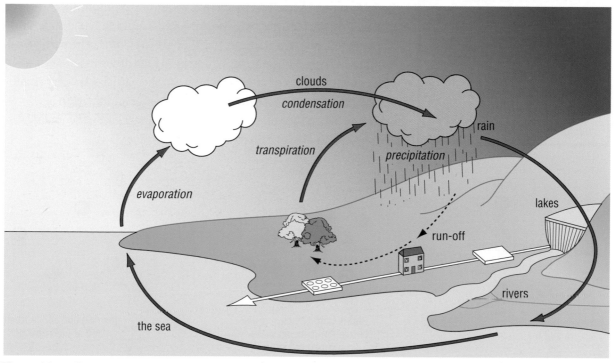

clouds

condensation

transpiration

rain

precipitation

evaporation

run-off

lakes

rivers

the sea

Figure 14.7   The water cycle.

4 Water vapour can also condense on the ground at night. What is this condensation called? If this substance freezes we give it another name. What is it?

5 Make a diagram to show the states of matter and the processes that change them. Start by copying out Figure 14.8 then add the words evaporating, melting, boiling, condensing, subliming and freezing to the appropriate arrows.

Water turns into a gas called water vapour by evaporation at any water surface. In the cool, upper air the water vapour condenses to form millions of water droplets that make the clouds. At the tops of the clouds it is so cold that the droplets freeze and form snowflakes. They fall through the cloud and melt to form raindrops. Falling water in the form of rain, snow or hail is called precipitation. Plant roots take up the water that passes through the soil, and their leaves return water to the atmosphere by transpiration.

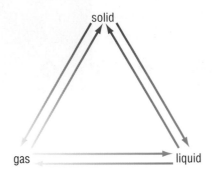

Figure 14.8   The interaction of the states of matter.

Particles of matter

Observations on the three states of matter and how they change can be explained by considering that matter is made of particles. This is called the 'particle theory of matter'.

Particles in the three states of matter

In solids, strong forces hold the particles together in a three-dimensional structure. In many solids the particles form an orderly arrangement called a lattice. The particles in all solids move a little. They do not change position but vibrate to and fro about one position.

In liquids, the particles have more energy than in the solid state, and the forces that hold the particles together are overcome. The particles in a liquid can change position by sliding over each other.

In gases, the forces of attraction between the particles are very small and the particles can move away from each other and travel in all directions. When they hit each other or the surface of their container they bounce and change direction.

6 According to the particle theory, why do liquids flow but solids do not?

7 How is the movement of particles in gases different from the movement of particles in liquids?

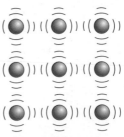

solid particles vibrate to and fro

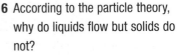

liquid particles have some freedom and can move around each other

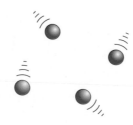

gas particles move freely and at high speed

Figure 14.9   Arrangement of particles in a solid, a liquid and a gas.

Particles and changes of state

Expanding and melting

If a solid is heated, it expands and then melts. The heat provides the particles with more energy. The energy makes the particles vibrate more strongly and push each other a little further apart – the solid expands. If the solid is heated further, the energy makes the particles vibrate so strongly that they slide over each other and become a liquid. During the time from when the solid starts to melt until it has completely turned into a liquid its temperature does not rise. All the heat energy is used to separate the particles so that they can flow over one another.

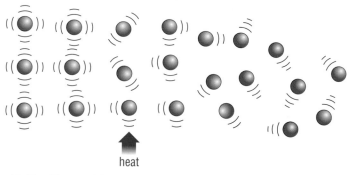

heat

Figure 14.10 The particle arrangement in a solid changes as the heat turns it into a liquid.

Figure 14.11 The water of this waterfall has frozen to form ice.

Freezing

If a liquid is cooled sufficiently the particles lose so much energy that they can no longer slide over each other. The only movement possible is the vibration to and fro about one position in the lattice. The liquid has become a solid.

Evaporation

The particles in a liquid have different amounts of energy. The particles with the most energy move the fastest. High energy liquid particles near the surface move so fast that they can break through the surface and escape into the air and form a gas.

Boiling

When a liquid is heated all the particles receive more energy and move more quickly. The fastest moving particles escape from the liquid surface or collect in the liquid to form bubbles. The bubbles rise to the surface and burst open into the air. The fast moving particles released from the liquid form a gas.

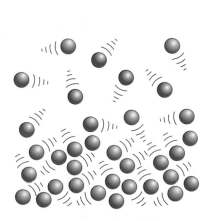

Figure 14.12 Evaporation.

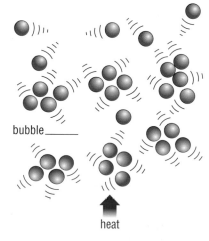

bubble

heat

Figure 14.13 Boiling.

Figure 14.14 Breathing onto a cold window causes water vapour in your breath to condense.

8 How is melting different from evaporation?

9 How is boiling different from sublimation?

10 How are condensation and freezing similar?

Condensation

The particles in a gas possess a large amount of energy which they use to move. If the particles are cooled they lose some of their energy and slow down. If the gas is cooled sufficiently, the particles lose so much energy that they can no longer bounce off each other when they meet. The particles now slide over each other and form a liquid.

Sublimation

When substances such as solid carbon dioxide or solid iodine sublime on heating, the energy the particles receive makes them separate and form a gas without forming a liquid first.

Figure 14.15 When solid iodine is heated it sublimes to form a gas. This is always done in a fume cupboard.

Pressure and changes of state

The state of matter of a substance can be changed by changing the pressure acting on it. Under very high pressure a gas can be turned into a liquid or a liquid into a solid.

Pressure on ice

As skaters move across the ice their weight pushes down through the small surface of the blades and makes a large pressure. The ice beneath the blade melts. When the skaters have passed by, the pressure is reduced on the ice surface and the water there freezes again. This change happens because ice is less dense than its liquid form, water. The change does not happen with other solids because they are denser than the liquids they form when they melt.

On other planets

The conditions on other planets in the Solar System are very different from conditions on Earth. Jupiter is made from a very large amount of hydrogen. The pressures and temperatures near the centre of the planet have made the hydrogen there into a solid like a metal. Above the solid hydrogen there is a vast ocean of liquid hydrogen.

Most of the planet Uranus is made from ammonia (the strong smelling gas that sometimes evaporates from a baby's nappy), methane (the gas used in cookers and fires) and water. Near the centre of the planet these substances are solids or liquids.

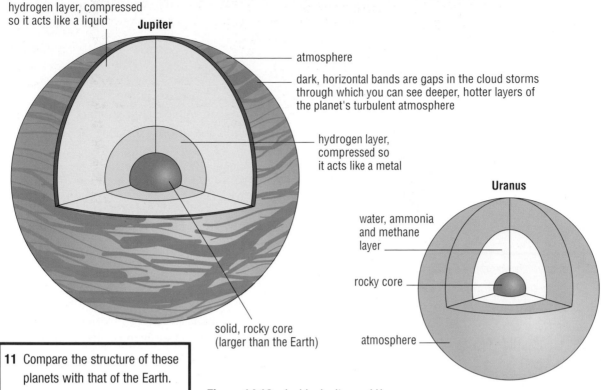

hydrogen layer, compressed
so it acts like a liquid

Jupiter

atmosphere

dark, horizontal bands are gaps in the cloud storms
through which you can see deeper, hotter layers of
the planet's turbulent atmosphere

hydrogen layer,
compressed so
it acts like a metal

Uranus

water, ammonia
and methane
layer

rocky core

atmosphere

solid, rocky core
(larger than the Earth)

11 Compare the structure of these
planets with that of the Earth.

Figure 14.16 Inside Jupiter and Uranus.

Atmospheric pressure on Earth

The atmosphere is a mixture of gases that covers the surface of the Earth.
The atmosphere is 1000 km thick and pushes on every square centimetre
of the Earth's surface. The pressure of the atmosphere at sea level is called
standard pressure and is about 10 N/cm². It is the pressure at which the
boiling point of any substance is measured. At the top of very high
mountains the pressure of the atmosphere is less than at sea level.

Boiling and low pressure

If a flask is connected to a vacuum pump and some of the air is sucked
out there is less air inside the flask to push on the surfaces and the air
pressure is smaller. The reduced air pressure allows evaporation to take

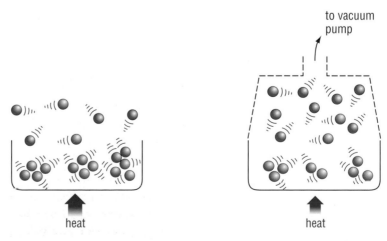

to vacuum
pump

heat

heat

Figure 14.17 Lowering the pressure lowers the boiling point of a liquid.

place more quickly and less heat is needed to make the liquid boil. Lowering the atmospheric pressure on a liquid lowers the boiling point of the liquid.

Boiling and high pressure

When a gas gets hot it expands and increases its pressure on the surfaces around it. If water is boiled in a pan with a lid, the steam escaping from the water pushes on the lid and makes it rise – allowing the gas to escape.

Diffusion

Diffusion is a process in which one substance spreads out through another. It occurs in liquids and gases. For example, if you put a drop of ink in a beaker of water the ink spreads out through the water by diffusion and colours it. The gases escaping from food cooking in the kitchen can move by diffusion to other rooms in the home. The moving particles in the different liquids flow over each other and the particles in the different gases bounce off each other. These movements eventually spread all the particles of one substance evenly through the other.

Liquids are denser than gases and this makes diffusion in liquids much slower than diffusion in gases.

13 Draw particle diagrams similar to those on page 135 to show the following processes:
a) sublimation,
b) condensation,
c) diffusion.

At start

After an hour

After a day

Figure 14.18   Black ink diffusing through a beaker of water.

Testing for purity

If water contains other substances dissolved in it, the water is impure. The impure water forms ice that melts at a temperature below 0 °C and boils at a temperature above 100 °C. The melting and boiling points of a substance can therefore also be used to find out if a substance in the laboratory is pure.

Summary

- There are three kinds or states of matter. They are solid, liquid and gas (*see page 130*).
- Each state of matter has properties that are different from other states (*see pages 130–131*).
- The properties of matter lead to different uses (*see page 131*).
- Matter can be changed from one state to another by the processes of melting, freezing, evaporation, boiling, condensation and sublimation (*see page 131*).
- The particle theory of matter can be used to explain how matter behaves (*see page 134*).
- The particles in the three states of matter behave differently (*see page 134*).
- When the activity of the particles changes, matter changes from one state to another (*see page 135*).
- The state of matter can be changed by changing the pressure acting on the substance (*see page 136*).
- Diffusion is a process in which one substance spreads out through another (*see page 138*).
- The melting and boiling points of a substance can be used to test for the purity of the substance (*see page 138*).

End of chapter questions

1 How many different materials is your shoe made from? What are the properties of each material? How are these properties useful?

2 Use the particle theory of matter to explain what happens to the particles when an ice cube melts and the water it produces evaporates.

15 Acids and bases

Acids

corrosive

Figure 15.1 The hazard symbol for a corrosive substance.

Most people think of acids as corrosive liquids which fizz when they come into contact with solids and burn when they touch the skin. This description is true for many acids and when they are being transported the container holding them has the hazard symbol shown in Figure 15.1.

Some acids are not corrosive and are found in our food. They give some foods their sour taste. This property gave acids their name. The word acid comes from the Latin word *acidus* meaning sour.

Many acids are found in living things. Tables 15.1 and 15.2 show some acids found in plants and animals.

Table 15.1 Acids found in plants.

Acid	Plant origin
citric acid	orange and lemon juice
tartaric acid	grapes
ascorbic acid	vitamin C in citrus fruits and blackcurrants
methanoic acid	nettle sting

Table 15.2 Acids found in animals.

Acid	Animal origin
hydrochloric acid	human stomach
lactic acid	muscles during vigorous exercise
uric acid	urine, excretory product from DNA in food
methanoic acid	ant sting

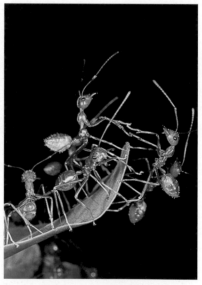

Figure 15.2 Animals and plants that produce acid.

The acid in vinegar

Ethanoic acid is found in vinegar and is produced as wine becomes sour. The wine contains ethanol produced by fermentation, and also has some oxygen dissolved in it from the air. Over a period of time, the oxygen reacts with the ethanol and converts it to ethanoic acid. This is an oxidation reaction and the reaction happens more quickly if the wine bottle is left uncorked.

1 Why does wine go sour faster if the cork is removed from the bottle?

Organic acids and mineral acids

The acids produced by plants and animals (with the exception of hydrochloric acid) are known as organic acids. Ethanoic acid is an organic acid and was the first to be used in experiments. Over the period AD750–1600 the mineral acids were discovered by alchemists. The first mineral acid to be discovered was nitric acid. It was used to separate silver and gold. When the acid was applied to a mixture of the two metals it dissolved the silver but not the gold. Later, sulphuric acid and then hydrochloric acid were discovered. These mineral acids are much stronger than ethanoic acid and allow more chemical reactions to occur. The use of these acids led to many chemical discoveries.

Figure 15.3   Bottles of dilute and concentrated acids.

2 How do you think the terms
   a) organic acids and
   b) mineral acids came to be used?
3 Acids in the laboratory are stored in labelled bottles as shown in Figure 15.3.
   a) Which acids are dilute and which are concentrated?
   b) How is a dilute solution different from a concentrated one?

A model volcano

In the past you may have made a model volcano. To do this you may have added a tablespoon of baking soda to an empty plastic drink bottle and then built a mound of sand around the bottle so that it looked like a conical volcano. Finally you may have added red dye to half a cup of vinegar then poured the vinegar into the bottle. Moments later a red froth would have emerged from the top of the bottle and flowed down the cone of sand, like lava flowing down a volcano (see Figure 15.4, overleaf).

Although the model looks impressive it does not illustrate how lava is formed, but it does show the power of a chemical reaction. Vinegar is an acid but if you were to test the 'lava' for acidity (see page 143) you might not find any. The chemical reaction has neutralised the acid. It is called a neutralisation reaction. A substance which neutralises an acid, like the baking soda, is called a base.

Figure 15.4   The ingredients (left) for making a model volcano (right).

Bases

As bases neutralise acids they are sometimes described as having properties which are opposite to acids. Bases are metal oxides, hydroxides, carbonates or hydrogencarbonates.

Some bases are soluble in water. They are called alkalis. Sodium hydroxide and potassium hydroxide are examples of alkalis that are used in laboratories. When they dissolve they form alkaline solutions. A concentrated alkaline solution is corrosive and can burn the skin. The same hazard symbol as the one used for acids (see Figure 15.1, page 140) is used on containers of alkalis when they are transported.

Even dilute solutions of alkali such as dilute sodium hydroxide solution react with fat on the surface of the skin and change it into substances found in soap. Many household cleaners used for cleaning metal, floors and ovens contain alkalis and must be handled with great care.

4 Which of the following substances are bases – copper chloride, sodium hydroxide, calcium carbonate, magnesium sulphate, copper oxide, lead nitrate, sodium hydrogencarbonate?

Figure 15.5   Alkalis used in the home.

Figure 15.6 Pink and blue hydrangeas.

---

5 Why are bases sometimes described as the opposite of acids?

6 How are acids and bases similar?

---

7 Here are some measurements of solutions that were made using a pH meter:

A 0, B 11, C 6, D 3, E 13, F 8.

a) Which of the solutions are
   i) acids,
   ii) alkalis?

b) If the solutions were tested with universal indicator paper, what colour would the indicator paper be with each one? (See Figure 15.8.)

c) Fresh milk has a pH of 6. How do you think the pH would change as it became sour? Explain your answer.

Detecting acids and alkalis

Some substances change colour when an acid or an alkali is added to them. Litmus is a substance which is extracted from a living organism called lichen. In chemistry it is used as a solution or is absorbed onto paper strips. Litmus solution is purple but it turns red when it comes into contact with an acid. Litmus paper for testing for acids is blue. The paper turns red when it is dipped in acid or a drop of acid is put on it. When an alkali comes into contact with purple litmus solution the solution turns blue. Litmus paper used for testing for an alkali is red. When red litmus paper comes into contact with an alkali it turns blue.

Hydrangeas have pink flowers when they are grown in a soil containing lime and blue flowers when grown in a lime-free soil. The colour of the flowers can be used to assess the alkalinity of the soil.

Universal indicator (see below) turns red, pink, orange, yellow, blue and purple when it comes into contact with an acid or alkali. The colour shows how weak or strong the acid or alkali is.

Strong and weak acids and alkalis

The strength of an acid or alkali does not describe whether the solution is dilute or concentrated. The strength is measured on the pH scale (see Figure 15.8, overleaf). On this scale the strongest acid is 0 and the strongest alkali is 14. A solution with a pH of 7 is neutral. It is neither an acid nor an alkali. An electrical instrument called a pH meter is used to measure the pH of an acid or alkali accurately.

Figure 15.7 A pH meter in use.

For general laboratory use, the pH of an acid or an alkali is measured with universal indicator. This is made from a mixture of indicators. Each indicator changes colour over part of the range of the pH scale. By combining the indicators together, a solution is made that gives particular colours over the whole of the pH range. These are shown in Figure 15.8.

8 Here are some results of solutions tested with universal indicator paper:
sulphuric acid – red, •
metal polish – dark blue,
washing-up liquid – yellow, •
milk of magnesia – light blue, •
oven cleaner – purple,
car battery acid – pink. •
Arrange the solutions in order of their pH, starting with the one with the lowest pH.

9 Identify the strong and weak acids and alkalis from the results shown in questions 7 and 8.

10 Look at page 141 about acids and predict whether nitric acid is a strong or a weak acid. Explain your answer.

11 A sample of acid rain turned universal indicator yellow. What would you expect its pH to be? Is it a strong or a weak acid?

12 Write word equations for the reactions between
a) sulphuric acid and zinc oxide,
b) hydrochloric acid and calcium hydroxide,
c) nitric acid and calcium carbonate.

13 How is the neutralisation of a carbonate different from the neutralisation of an oxide or a hydroxide?

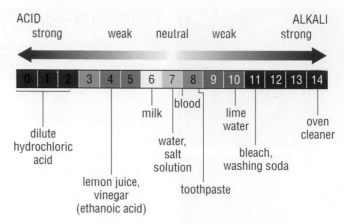

Figure 15.8 The pH scale showing universal indicator colours.

Neutralisation

When an acid reacts with a base a process called neutralisation occurs in which a salt and water are formed. This reaction can be written as a general word equation:

acid + base → salt + water

Specific examples of neutralisation reactions are:

sulphuric + magnesium → magnesium + water
acid   oxide   sulphate

hydrochloric + sodium → sodium + water
acid   hydroxide   chloride

hydrochloric + zinc → zinc   + water + carbon
acid   carbonate   chloride   dioxide

nitric + sodium → sodium + water + carbon
acid   hydrogencarbonate   nitrate   dioxide

Using neutralisation reactions

When you are stung by a nettle, the burning sensation on your skin is caused by methanoic acid. You can neutralise the acid by rubbing a dock leaf on the wound. As you press the dock leaf against the wound, a base in the leaf juices reacts with the acid in the sting and neutralises it so the burning sensation stops.

A bee sting is acidic and may be neutralised by soap, which is an alkali. A wasp sting is alkaline and may be neutralised with vinegar, which is a weak acid.

Sometimes the stomach produces too much acid, which causes indigestion. The acid is neutralised by taking a tablet containing either magnesium hydroxide, calcium carbonate, aluminium hydroxide or sodium hydrogencarbonate (see Figure 15.9, opposite).

Lime is used to neutralise acidity in soil. When it is applied to fields it makes them appear temporarily white, as Figure 15.10 shows.

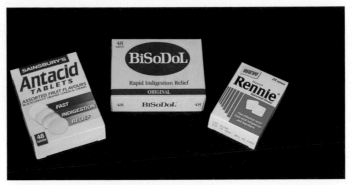

Figure 15.9 A selection of tablets which neutralise the acid that causes indigestion.

Figure 15.10 Liming fields to improve crop production.

The soda–acid fire extinguisher (Figure 15.11) contains a bottle of sulphuric acid and a solution of sodium hydrogencarbonate. When the plunger is struck or the extinguisher is turned upside down, the acid mixes with the sodium hydrogencarbonate solution and a neutralisation reaction takes place. The pressure of the carbon dioxide produced in the reaction pushes the water out of the extinguisher and onto the fire.

Figure 15.11 A soda–acid fire extinguisher.

Summary

- Some acids are made by living things (*see page 140*).
- Ethanoic acid in vinegar is made by the oxidation of ethanol in wine (*see page 140*).
- The mineral acids are nitric acid, sulphuric acid and hydrochloric acid (*see page 141*).
- Bases are metal oxides, hydroxides, carbonates and hydrogencarbonates (*see page 142*).
- Bases that dissolve in water are called alkalis (*see page 142*).
- An acid can be detected by its ability to turn blue litmus paper red (*see page 143*).
- An alkali can be detected by its ability to turn red litmus paper blue (*see page 143*).
- The pH scale is used to measure the degree of acidity or alkalinity of a liquid (*see page 143*).
- When an acid reacts with a base a neutralisation reaction takes place (*see page 144*).
- Neutralisation reactions have a wide range of uses (*see page 144*).

End of chapter questions

1 Write an account entitled 'The acids in our lives'.

2 How can you tell when an acid has neutralised an alkali?

16 Simple chemical reactions

A chemical reaction took place in the model volcano (see page 142) to make the 'lava' flow. When substances take part in a chemical reaction one or more new substances are created. One new substance was easily detected in the chemical reaction in the volcano. It was a gas that made a huge number of bubbles in the 'lava'.

Reactants and products

The substances that take part in the reaction are called the reactants. The substances that form as a result of the chemical reaction are called the products:

reactants → products

Chemists use chemical equations to describe the chemical reactions. They save time and space and provide the essential information about the reaction in an easy-to-read form. The simplest chemical equations are word equations.

In an equation the reactants are written on the left hand side and the products on the right hand side. If two or more reactants or products are featured in the equation they are linked together by plus (+) signs:

reactant A + reactant B → product C + product D

An arrow points from the reactants to the products. Most reactions are not reversible and there is only one arrow.

Some reactions are reversible (they can go in either direction) and a special arrow sign points in both directions:

$$A + B \rightleftharpoons C + D$$

1 What is the difference between a product and a reactant in a chemical reaction?
2 How can you tell from the equation if the reaction is reversible or not?

Test for hydrogen

Hydrogen is a colourless gas that does not smell. It can be detected by the following test. Hold a small test-tube of hydrogen upside down and remove the bung. Hold a lighted splint below the open mouth of the test-tube and a popping sound will be heard. The hydrogen in the tube combines with the oxygen in the air and this explosive reaction makes the popping sound. The word equation for the reaction is:

hydrogen + oxygen → hydrogen oxide

The common name for hydrogen oxide is water, and this is the word that is normally used in word equations:

hydrogen + oxygen → water

Acids and metals

Some metals react with acids and produce a salt and hydrogen. The term salt is used in everyday language for the compound sodium chloride. In chemistry, it can mean any metal compound made from an acid. The general word equation for the reaction between a metal and an acid is:

metal + acid → salt + hydrogen

An example of this is the reaction between zinc and hydrochloric acid. The word equation for this reaction is:

zinc + hydrochloric acid → zinc chloride + hydrogen

Figure 16.1 shows the apparatus and reactants set up to demonstrate this reaction.

3 Bubbles of hydrogen are released from the surface of the acid and build up a high gas pressure in the flask. What do you think happens next to
   a) the hydrogen in the flask,
   b) the water in the test-tube?
4 Write a word equation for the reaction between magnesium and sulphuric acid.

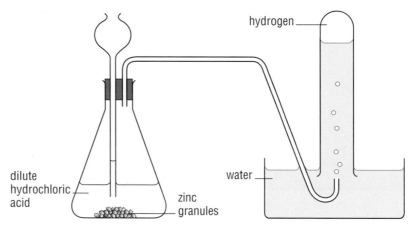

Figure 16.1   Apparatus for the collection of hydrogen.

Test for carbon dioxide

Lime water is a dilute solution of calcium hydroxide. It is used to test for carbon dioxide gas. If a gas is thought to be carbon dioxide, it is bubbled through lime water. If carbon dioxide is present, a chemical reaction takes place in which calcium carbonate is made. This white substance is insoluble in water and forms a white precipitate which makes the lime water cloudy or milky.

Clear lime water

Carbon dioxide is bubbled through

Lime water turns cloudy

Figure 16.2   A test for carbon dioxide.

Acids and carbonates

A carbonate is a compound that contains carbon and oxygen combined together. When a carbonate reacts with an acid, carbon dioxide is released and a salt and water form. This reaction may be written as the general word equation:

carbonate + acid → carbon dioxide + salt + water

An example of this is the reaction between magnesium carbonate and hydrochloric acid. The word equation for this reaction is:

magnesium + hydrochloric → magnesium + water + carbon
carbonate   acid   chloride   dioxide

5 Write the word equation for the reaction between sulphuric acid and copper carbonate.

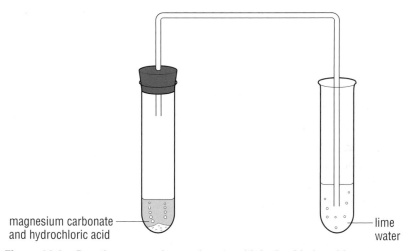

magnesium carbonate
and hydrochloric acid

lime
water

Figure 16.3 Reacting magnesium carbonate with hydrochloric acid.

Combustion

Combustion is a chemical reaction in which energy is given out as heat. If a flame develops, combustion is then called burning. In burning, energy is also given out as light and sound.

Figure 16.4 Burning needs to take place in air (or oxygen).

Burning

Many substances are burned to provide heat or light. They are called fuels. Wood, coal, coke, charcoal, oil, diesel oil, petrol, natural gas and wax are examples of fuels. The heat may be used to warm buildings, cook meals, make chemicals in industry, expand gases in vehicle engines and turn water into steam to drive generators in power stations. Some gases and waxes are used to provide light in caravans and tents.

6 Give a use for each of the fuels listed in the paragraph on burning. How many different uses can you find?

Natural gas is an example of a hydrocarbon. It is made of carbon and hydrogen. When natural gas burns, carbon dioxide and water (hydrogen oxide) are produced. Many other fuels such as coal, coke and petrol contain hydrocarbons.

Investigating a burning candle

A candle can be used to investigate how fuels burn.

Investigation 1

If a burning candle is put under a thistle funnel which is attached to the apparatus shown in Figure 16.5 and the suction pump is switched on, a liquid collects in the U-tube and the lime water turns cloudy.

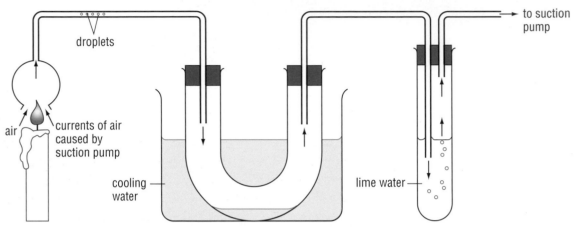

Figure 16.5   Testing the products of a burning candle.

When the liquid is tested with cobalt chloride paper, the paper turns from blue to pink. This shows that the liquid is water. The cloudiness in the lime water indicates that carbon dioxide has passed into it.

Investigation 2

If a beaker is placed over a burning candle, the candle will burn for a while and then go out. A change has taken place in the air that makes it incapable of letting things burn in it.

The test for oxygen is made by plunging a glowing piece of wood into the gas being tested. If the gas is oxygen, the wood bursts into flame. When air from around the burned-out candle is tested for oxygen, the glowing wood goes out. This indicates that oxygen is no longer present. The oxygen in the air under the beaker has been used up by the burning candle.

From the information provided by these two investigations with candles, the following word equation can be set out:

hydrocarbon + oxygen → carbon dioxide + water + *energy*

Natural gas is a hydrocarbon called methane. When it burns, it breaks down exactly like the hydrocarbons in candle wax. The word equation for this reaction is:

methane + oxygen → carbon dioxide + water + *energy*

Both of these word equations are examples of complete combustion. This only happens when there is enough oxygen available.

The Bunsen burner

The Bunsen burner uses natural gas as a fuel. The parts of a Bunsen burner are shown in Figure 16.6.

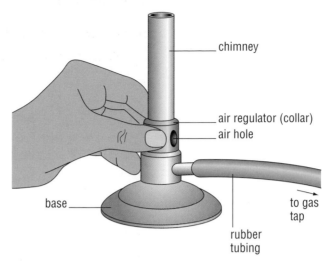

Figure 16.6 The Bunsen burner.

The air regulator or collar must be turned to fully close the air hole before the burner is lit. The match should be lit and placed to one side of the top of the chimney before the gas tap is switched on.

When the gas is switched on, it shoots out through the jet and up the chimney. Not all the carbon in the gas combines with the oxygen straight away and carbon particles are produced. These are heated and give out a yellow light which makes the flame. This flame should not be used to heat anything, as the carbon particles form soot on the surface of the apparatus being heated. The flame produced with the air hole closed is called a luminous (or wavy) flame. It is silent. The carbon in the flame reacts with oxygen in the air and forms carbon dioxide.

If the collar is turned and the air hole is fully opened, air mixes with the gas in the chimney. The gases rush up the chimney and form a blue cone of unburnt gas at the top of the chimney. Above the cone, the complete combustion of methane takes place. The flame made when the air hole is completely open is non-luminous and makes a roaring sound. The hottest part of the non-luminous flame is a few millimetres above the tip of the blue cone of unburnt gas.

Less heat energy is released by the luminous flame than the non-luminous flame because the carbon does not all react with oxygen at once. When the Bunsen burner is not being used for heating, the air hole should be closed to give a luminous flame, as a safety precaution against burns.

The size of the flame is controlled by the gas tap on the bench. If the tap is fully open a large flame is produced. A smaller flame is produced by partially closing the gas tap.

7 Why is one flame hotter than the other?

8 Why does closing the gas tap a little reduce the size of the flame?

9 What safety precautions should you take when using a Bunsen burner? Explain the reason for each precaution.

10 When solid zinc oxide is placed in an aqueous solution of hydrochloric acid, zinc chloride is produced which dissolves in the water. Water is also produced.
   a) Write the word equation for this reaction.
   b) Write in the state symbols.

State symbols

The chemicals taking part in a reaction and the products that they form may be in different states of matter. These states can be represented by symbols in the equation. In addition to (s) for solid, (l) for liquid and (g) for gas, there is a fourth symbol. It is (aq) and shows that the chemical is in an aqueous solution, which means that it is dissolved in water. The symbols are added after the name of each chemical. For example:

calcium + hydrochloric → calcium + water(l) + carbon
carbonate(s)   acid(aq)   chloride(aq)   dioxide(g)

Summary

- An equation for a chemical reaction features reactants and products separated by an arrow (*see page 147*).
- When a lighted splint is held below the open mouth of a test-tube filled with hydrogen, a popping sound is heard (*see page 147*).
- An acid reacts with a metal to produce a salt and hydrogen (*see page 148*).
- Carbon dioxide gas turns lime water milky (*see page 148*).
- An acid reacts with a carbonate to produce a salt, water and carbon dioxide (*see page 149*).
- Burning is a type of combustion in which a flame is produced (*see page 149*).
- The Bunsen burner is a device which allows the combustion of methane to be controlled to supply heat for experiments (*see page 151*).
- State symbols can be added to equations to provide information about the states of the reactants and products (*see above*).

End of chapter questions

1 When an acid is poured onto a metal, bubbles of gas appear.
   a) What could the gas be?
   b) How would you perform an investigation to test your idea?

2 Two reactants in a flask are producing bubbles of gas that turn lime water milky.
   a) What could the reactants in the flask be?
   b) If the liquid reactant was poured onto a metal what would be produced?

3 When you light a Bunsen burner, what happens to the methane that flows into it?

17 Separation of mixtures

A mixture is composed of two or more separate substances. The composition of a mixture may vary widely. One mixture of two substances, A and B, might have a large amount of A and a small amount of B. Another mixture might have a small amount of A and a large amount of B.

A =

B =

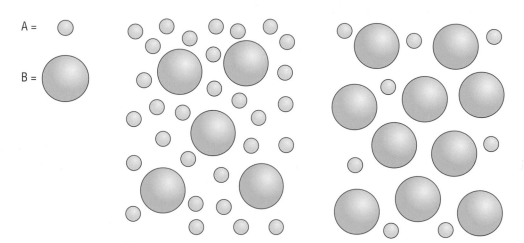

Figure 17.1 Two different mixtures of A and B.

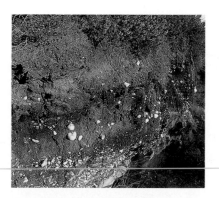

Figure 17.2 A cross-section of soil.

The substances in a mixture can be separated by physically removing one substance from another, as shown in the separating techniques in this chapter.

Different kinds of mixtures

Solid/solid mixtures

Soil is a mixture of different solid particles. Some particles such as clay are very small while others such as sand are larger.

Solid/liquid mixtures

If clay is stirred with water it forms a cloudy mixture. The tiny clay particles are suspended in the water. The mixture is called a suspension. If a solid dissolves in a liquid a solution is made.

Solid/gas mixtures

The smoke rising in the hot air from a bonfire contains particles of soot and ash. A mixture of solid and gas also occurs in the dust produced when the wall of a building is being cleaned by blasting a jet of sand at it.

Liquid/liquid mixtures

Milk is a mixture of tiny droplets of fatty oil in water (see Figure 17.3). This kind of mixture is called an emulsion. An emulsion mixture is also found in some kinds of paint.

Figure 17.3 Fat globules in whole milk, as seen under a high power microscope.

Gas/gas mixtures

Gases can move freely and when two different gases meet they mix. The most common mixture of gases is the one around you right now – the air.

Liquid/gas mixtures

When water vapour in the air condenses above the cool surface of a lake or a field, the tiny droplets form a mist or fog. When you press the top of an aerosol can, for example, a mist of liquid droplets in a gas (an 'aerosol') is sprayed into the air.

Gas/liquid mixtures

When bubbles of gas are trapped in a liquid they form a foam. Foam is made when the nozzle of a shaving foam cylinder is pressed. Some products that protect you from sunburn are foams.

Solutions

The most common form of a mixture in chemical experiments is the solution. A solution is made when a substance, called a solute, mixes with a liquid, called a solvent, in such a way that the solute can no longer be seen. This type of mixing is called dissolving.

Although the solute cannot be seen it has not taken part in a chemical reaction and can be recovered from the solution by separating it from the solvent. The solute may be a solid, liquid or gas.

Copper sulphate dissolves in water to form a blue solution.

Clay does not dissolve in water, but forms a suspension that settles to the bottom after some time.

Figure 17.4 Soluble and insoluble substances.

1 What is the difference between a solvent and a solute?
2 What is the difference between a substance that is soluble in water and one that is insoluble in water?
3 What is the difference between an immiscible substance and an emulsion?

A liquid that dissolves in a solvent, water for example, is said to be miscible with water. A liquid that does not dissolve in a solvent is said to be immiscible with it.

A gas or a solid that dissolves in a solvent is said to be soluble in that solvent. A solid or gas that does not dissolve in a solvent is said to be insoluble in that solvent (see Figure 17.4).

Saturated solutions

If the temperature of a solvent is kept constant, and the solute is added in small amounts, there comes a time when no more solute will dissolve. The solution is then said to be saturated. If the temperature of the saturated solution is raised, it is able to take in more solute until it becomes saturated at the new temperature.

Solubility

The solubility of a solute in a solvent at a particular temperature is the maximum mass of the solute that will dissolve in 100 g of the solvent, before the solution becomes saturated.

If the temperature of the solvent is raised the solubility of the solute usually increases (see Figure 17.5). If the solubilities of a substance at different temperatures of the solvent are plotted on a graph, a solubility curve is made.

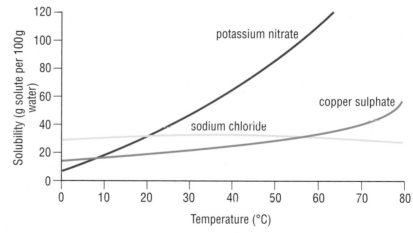

Figure 17.5 Solubility curves.

4 What do the solubility curves of the three substances in Figure 17.5 show?
5 How does the solubility of potassium nitrate change when the temperature of the solvent is raised from 30 to 50 °C?

Solubility of gases

More gas will dissolve in cold water than in warm water – this is the opposite of what happens with solids. Hot water entering a river from a power station can warm the river water so much that not enough oxygen can dissolve in it for the fish to breathe. Some species of water animals can only live where there is a high concentration of oxygen in the water. These species must live in the cool waters of mountain streams.

6 How does the solubility of oxygen in water vary with the temperature of the water?

Different solvents

Water has been called the universal solvent because so many different substances dissolve in it. However, there are many liquids used as solvents in a wide range of products. Ethanol is used in perfumes, aftershaves and glues. Propanone is used to remove nail varnish and grease. Gloss paint is dissolved in white spirit.

Substances that dissolve in one solvent do not necessarily dissolve in others. Salt dissolves in water but not in ethanol. White sugar dissolves in both.

Separating mixtures

The substances in a mixture have not taken part in a chemical reaction and have kept their original characteristics. These characteristics are used in the following techniques to separate the substances.

Separating a solid/solid mixture

A mixture of two solids with particles of different sizes may be separated by using a sieve. The particles in soil are analysed by putting the soil in the top compartment of a soil sieve and shaking it. Each part of the soil sieve has a mesh with smaller holes than the one before. Different sized particles are caught in each layer as the soil moves from the top to the bottom.

Magnetic materials can be separated from non-magnetic materials by passing the materials close to a magnet. In a metal separator, the cylinder is a magnet that attracts iron and steel items to it, while other metals fall away. The magnetic materials are then knocked off the cylinder into another collection bay, as shown in Figure 17.7.

Many metals are found in rocks combined with other substances and with a great deal of worthless material. This mixture is known as an ore.

Metal compounds and the worthless material can be separated using a flotation cell. The ore is broken up into fragments and added to a mixture of water, oil and a range of chemicals. When compressed air is blown through the chemical mixture, a froth is produced which rises to

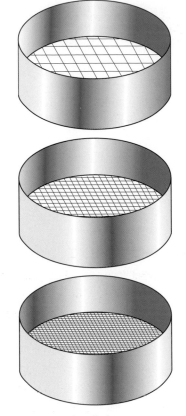

Figure 17.6 A soil sieve.

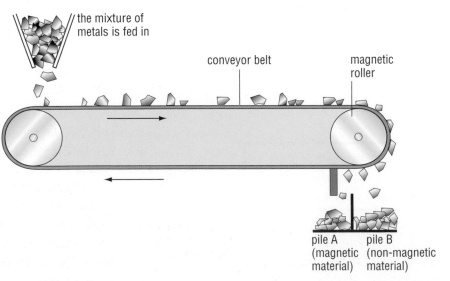

the mixture of metals is fed in

conveyor belt

magnetic roller

pile A (magnetic material)   pile B (non-magnetic material)

Figure 17.7 A magnetic separator.

7 What is used in a flotation cell to separate a metal compound out of its ore?

the surface. The chemicals also help the particles containing the metal to cling to the froth, while the worthless material is left behind. The skimmers at the surface remove the froth and valuable metal compound.

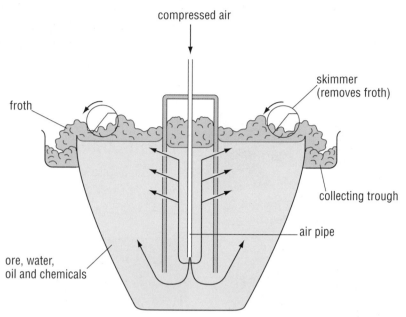

Figure 17.8 A flotation cell.

Figure 17.9 Decanting a liquid.

Separating an insoluble solid/liquid mixture

In the home, sieves are used to separate insoluble solids, such as peas, from liquids. This is possible because the particle size of the solid is very much larger than that of water. In chemistry, this is not usually the case and other methods are needed.

Large particles: decanting

Large particles of an insoluble solid in a liquid settle at the bottom of the liquid's container. They form a layer called a sediment. The liquid and solid can be separated by decanting. A liquid is decanted by carefully pouring it out of the container without disturbing the sediment at the bottom. At home, some medicines and sunburn lotions form a sediment in the bottom of the bottle and have to be shaken to mix the solid and liquid before being used.

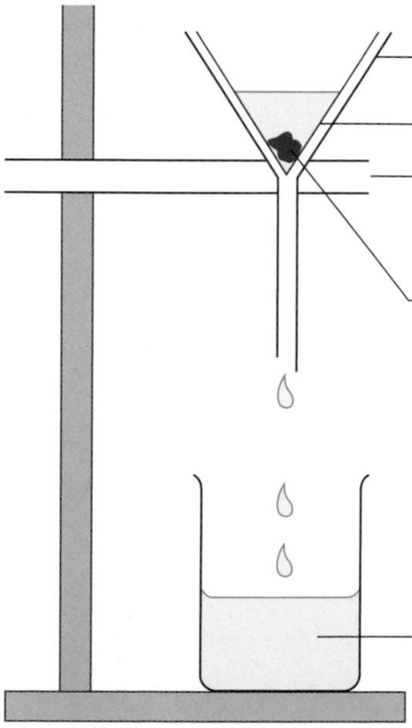

Small particles: filtration

In many laboratory experiments, filtration is carried out by folding a piece of filter paper to make a cone and inserting it in a filter funnel. The funnel is then supported above a collecting vessel and the mixture to be separated is poured into the funnel.

The filter paper is made of a mesh of fibres. It works like a sieve but the holes between the fibres are so small that only liquid can pass through them. The solid particles are left behind on the paper fibres.

filter funnel

filter paper

support

the solid remains in the filter as the residue

the liquid that filters through is called the filtrate

Figure 17.10 Filtration with a filter funnel.

Figure 17.11 Copper sulphate solution being evaporated over a water bath.

Separating a solute from a solute/solvent mixture

These methods can be used to separate a solid solute from a solvent.

Evaporation

If a solution is heated gently, the solvent evaporates from the surface until only the solid is left behind. Distilled water is made by boiling and condensing water to remove impurities.

Tap water and sea water may be compared with distilled water by setting up samples of all three kinds of water and heating them gently until all the liquid has evaporated, leaving only the solid content behind.

Crystallisation

A crystal is a solid structure with flat sides. Many substances form crystals. One way of making crystals is to start with a concentrated solution of a substance.

Figure 17.12 Crystals of copper sulphate formed in an evaporating dish.

As the solution is gently heated the solvent evaporates and the concentration of the solute in the solution rises until the solution is saturated (see page 155). If the heat is removed at this time and the saturated solution is left to cool, the solid will form crystals.

Separating several different solutes from a solvent: chromatography

A simple chromatography experiment can be performed with filter paper, a dropper, ink and water. A drop of ink is placed in the centre of the filter paper, then a drop of water is placed on top of it. The water dissolves the coloured pigments and spreads out through the filter paper, carrying the pigments with it. Each kind of pigment moves at a different speed to the others so that they spread out into different regions of the paper. When the separation is complete, the paper is dried. The paper with its separate pigments is called a chromatogram.

Substances that do not dissolve in water can be separated by chromatography by using other solvents, such as propanone. When one of these solvents is used, the chromatography paper is enclosed in a tank. This makes sure that the solvent vapour does not escape but surrounds the paper keeping it saturated with solvent and helps the substances to separate.

8 The unknown mixture used in Figure 17.13 is suspected of containing substances A, B and C. Do the results confirm this?

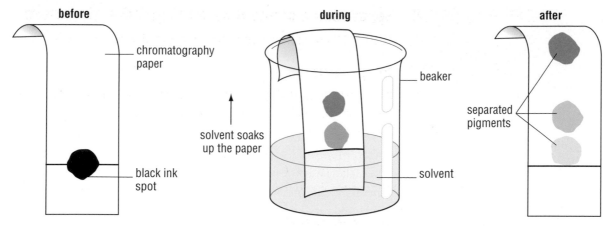

Figure 17.13 Simple paper chromatography.

Separating a solvent from a solute/solvent mixture

During evaporation or boiling, the liquid solvent is lost to the air. If the solvent is important, it can be separated from the mixture using a process called distillation.

In a very simple form of distillation, the solution is placed in a test-tube which is set up as shown in Figure 17.14.

The anti-bumping granules provide many places where bubbles of gas may form as the water boils. The bubbles are small and steadily rise to the liquid surface where they burst. Without the granules, fewer but larger bubbles form that rise and burst with such force that they shake the test-tube.

As the water boils the steam moves along the delivery tube. At first the tube is cool enough to make some of the steam condense but as more steam passes along the tube it becomes hotter and no more condensation takes place. The cold water in the beaker keeps the walls of the second test-tube cool so that most of the steam condenses there and

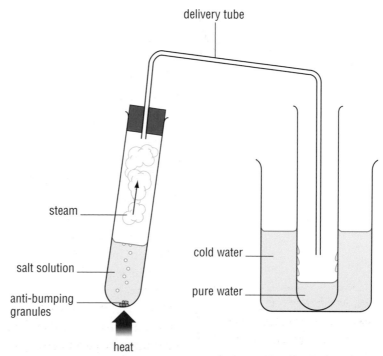

Figure 17.14 Simple distillation.

water collects at the bottom of the tube. The solid solute remains in the first tube. The purity of the water can be checked by boiling it and recording its boiling point with a thermometer.

Distillation with a Liebig condenser

The Liebig condenser is a glass tube surrounded by a glass chamber called a water jacket. During the distillation process, water is allowed to flow from the cold tap through the water jacket and down the sink. The water takes away the heat from the hot vapour in the tube of the condenser and causes condensation. The liquid formed by the condensed vapour is called the distillate. It flows down the tube and drips into the collection flask.

The condenser is named after Justus von Liebig (1803–1873), a famous chemist of the 19th century.

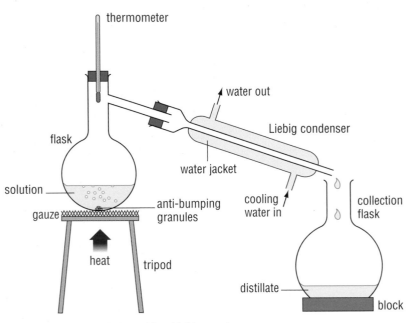

9 Why is the Liebig condenser more efficient than the simple distillation apparatus?

Figure 17.15   Distillation with a Liebig condenser.

Separating two miscible liquids

Fractional distillation

Two liquids with different boiling points, such as water (100 °C) and ethanol (78 °C), can be separated by fractional distillation.

The separation occurs in the fractionating column. This is filled with glass beads that provide a large surface area. During the fractionating process the liquids condense and evaporate from the surface many times. At first, in the lower part of the column, the water and ethanol vapours condense together on the cold beads. They warm them up and some of the ethanol and a little water vapour evaporate and move up a little further, then condense. As the solution continues to boil, this process of condensation and evaporation is repeated all the way up the column. Each time, more of the liquid with the lower boiling point – the ethanol – rises further, until it reaches the top. The ethanol vapour then passes down the Liebig condenser and is collected in the flask. As the vapour passes the thermometer, a temperature of 78 °C is recorded. This is the temperature at which pure ethanol boils. When most of the ethanol has passed the thermometer the temperature starts to rise. At this point, the process is stopped as the liquid in the flask is nearly all water.

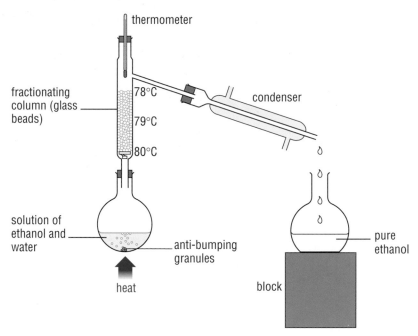

10 Why did the thermometer show 78 °C for some time?

11 If the distillation was left to run, what temperature would the thermometer rise to? Explain your answer.

12 How would you use the apparatus in Figure 17.16 to get a flask of ethanol and a flask of water?

Figure 17.16 The distillation of ethanol.

Separating immiscible liquids

When two immiscible liquids are mixed together they eventually form layers, if left to stand. This can be seen when oil and vinegar are mixed together to form salad dressing.

Figure 17.17 Salad dressing mixture after shaking (left) and after standing for 10 minutes (right).

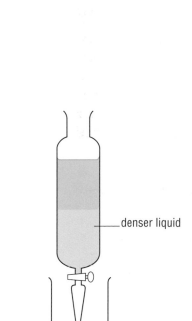

Figure 17.18 A separating funnel.

The less dense liquid forms a layer above the more dense liquid. The separating funnel (see Figure 17.18) can be used to separate them. The tap is opened to let the liquid in the lower layer flow away into a beaker. A second beaker can be used to collect the liquid from the upper layer.

Summary

- A mixture is composed of two or more substances. There are no fixed amounts in which they combine. They can be separated by physically removing one substance from the other (*see page 153*).
- There is a wide range of mixtures in which solids, liquids and gases combine together (*see page 153*).
- A solution is composed of a solute and a solvent (*see page 154*).
- When no more solute will dissolve in a solvent, the solution is said to be saturated (*see page 155*).
- The maximum mass of solute that will dissolve in 100 g of solvent, at a particular temperature, is known as the solubility of the substance at that temperature (*see page 155*).
- More gas will dissolve in cold water than in warm water (*see page 155*).
- There are many solvents. Substances that dissolve in one solvent may not dissolve in others (*see page 156*).
- Solids may be separated from each other with a sieve, a magnetic separator or a flotation cell (*see page 156*).
- Solids may be separated from liquids by decanting or by filtration (*see page 157*).
- A solid solute may be separated from a solvent by evaporation, crystallisation or chromatography (*see page 158*).
- A solvent may be separated from a solid solute by distillation (*see page 159*).
- Miscible liquids can be separated by fractional distillation (*see page 160*).
- Two immiscible liquids can be separated by a separating funnel (*see page 161*).

End of chapter question

1 How would you separate the different parts of a mixture of sand and salty water?

18 Elements

One of the main activities in chemistry is breaking down substances to discover what they are made of. During the course of this work chemists have discovered that some substances will not break down into simpler substances. These substances are called elements.

Discovery of the elements

Before 1669 the following elements had already been discovered – carbon, sulphur, iron, copper, arsenic, silver, tin, antimony, gold, mercury and lead. Some had been known for thousands of years, although they had not been recognised as elements. The order in which the other elements were discovered is shown in Table 18.1. This table uses mainly European historical data but it is known that the Chinese also practised alchemy, so some of the elements could have been discovered by them at an earlier date.

1 How many elements were discovered in
  a) the 17th century,
  b) the 18th century,
  c) the 19th century?
2 Which three scientists discovered the most elements?
3 How many Swedish scientists discovered new elements?
4 Which UK scientist discovered the most elements?

Table 18.1 The discovery of the elements.

Date	Element	Discoverer	Brief description
1669	Phosphorus	H. Brand (Germany)	white, red, black solid
1737	Cobalt	G. Brandt (Sweden)	reddish metal
1746	Zinc	A.S. Marggraf (Germany)	blue–white metal
1748	Platinum	A. de Ulloa (Spain)	blue–white metal
1751	Nickel	A.F. Cronstedt (Sweden)	silver–white metal
1753	Bismuth	C.F. Geoffroy (France)	silver–red metal
1766	Hydrogen	H. Cavendish (UK)	colourless gas
1771–1774	Oxygen	C.W. Scheele (Sweden) J. Priestley (UK)	colourless gas
1772	Nitrogen	D. Rutherford (UK)	colourless gas
1774	Chlorine	C.W. Scheele (Sweden)	green–yellow gas
1774	Manganese	J.G. Gahn (Sweden)	red–white metal
1781	Molybdenum	P.J. Hjelm (Sweden)	silver–grey metal
1783	Tellurium	F.J. Muller (Austria)	silver–grey solid
1783	Tungsten	J.J. de Elhuya, F. de Elhuya (Spain)	grey metal
1789	Zirconium	M.H. Klaproth (Germany)	shiny, white metal
1789	Uranium	M.H. Klaproth (Germany)	blue–white metal
1794	Yttrium	J. Gadolin (Finland)	shiny, grey metal
1795	Titanium	M.H. Klaproth (Germany)	silvery metal
1798	Beryllium	N-L Vauquelin (France)	brown powder
1798	Chromium	N-L Vauquelin (France)	silvery metal
1801	Niobium	C. Hatchett (UK)	grey metal
1802	Tantalum	A.G. Ekeberg (Sweden)	silvery metal

(continued)

Date	Element	Discoverer	Brief description
1803	Cerium	J.J. Berzelius, W. Hisinger (Sweden) M.H. Klaproth (Germany)	grey metal
1803	Palladium	W.H. Wollaston (UK)	silver–white metal
1804	Rhodium	W.H. Wollaston (UK)	grey–blue metal
1804	Osmium	S. Tennant (UK)	blue–grey metal
1804	Iridium	S. Tennant (UK)	silver–white metal
1807	Potassium	H. Davy (UK)	silver–white metal
1807	Sodium	H. Davy (UK)	silver–white metal
1808	Magnesium	H. Davy (UK)	silver–white metal
1808	Calcium	H. Davy (UK)	silver–white metal
1808	Strontium	H. Davy (UK)	silver–white metal
1808	Barium	H. Davy (UK)	silver–white metal
1811	Iodine	B. Courtois (France)	grey–black solid
1817	Lithium	J.A. Arfwedson (Sweden)	silver–white metal
1817	Cadmium	F. Stromeyer (Germany)	blue–white metal
1818	Selenium	J.J. Berzelius (Sweden)	grey solid
1824	Silicon	J.J. Berzelius (Sweden)	grey solid
1825–1827	Aluminium	H.C. Oersted (Denmark) F. Wohler (Germany)	silver–white metal
1826	Bromine	A.J. Balard (France)	red–brown liquid
1829	Thorium	J.J. Berzelius (Sweden)	grey metal
1830	Vanadium	N.G. Sefstrom (Sweden)	silver–grey metal
1839	Lanthanum	C.G. Mosander (Sweden)	metallic solid
1843	Terbium	C.G. Mosander (Sweden)	silvery metal
1843	Erbium	C.G. Mosander (Sweden)	silver–grey metal
1844	Ruthenium	K.K. Klaus (Estonia)	blue–white metal
1860	Caesium	R.W. Bunsen, G.R. Kirchhoff (Germany)	silver–white metal
1861	Rubidium	R.W. Bunsen, G.R. Kirchhoff (Germany)	silver–white metal
1861	Thallium	W. Crookes (UK)	blue–grey metal
1863	Indium	F. Reich, H.T. Richter (Germany)	blue–silver metal
1868	Helium	J.N. Lockyer (UK)	colourless gas
1875	Gallium	L. de Boisbaudran (France)	grey metal
1878	Ytterbium	J-C-G de Marignac (Switzerland)	silvery metal
1878–1879	Holmium	J.L. Soret (France) P.T. Cleve (Sweden)	silvery metal
1879	Scandium	L.F. Nilson (Sweden)	metallic solid
1879	Samarium	L. de Boisbaudran (France)	light grey metal
1879	Thulium	P.T. Cleve (Sweden)	metallic solid
1880	Gadolinium	J-C-G de Marignac (Switzerland)	silver–white metal

(continued)

Date	Element	Discoverer	Brief description
1885	Neodymium	C. Auer von Welsbach (Austria)	yellow–white metal
1885	Praseodymium	C. Auer von Welsbach (Austria)	silver–white metal
1886	Dysprosium	L. de Boisbaudran (France)	metallic solid
1886	Fluorine	H. Moissan (France)	green–yellow gas
1886	Germanium	C.A. Winkler (Germany)	grey–white metal
1894	Argon	W. Ramsay, Lord Rayleigh (UK)	colourless gas
1898	Krypton	W. Ramsay, M.W. Travers (UK)	colourless gas
1898	Neon	W, Ramsay, M. W. Travers (UK)	colourless gas
1898	Polonium	Mme M.S. Curie (Poland/France)	metallic solid
1898	Xenon	W. Ramsay, M.W. Travers (UK)	colourless gas
1898	Radium	P. Curie (France), Mme M.S. Curie (Poland/France), M.G. Bermont (France)	silvery metal
1899	Actinium	A. Debierne (France)	metallic solid
1900	Radon	F.E. Dorn (Germany)	colourless gas
1901	Europium	E.A. Demarçay (France)	grey metal
1907	Lutetium	G. Urbain (France)	metallic solid
1917	Protactinium	O. Hahn (Germany), Fr L. Meitner (Austria), F. Soddy, J.A. Cranston (UK)	silvery metal
1923	Hafnium	D. Coster (Netherlands) G.C. de Hevesy (Hungary/Sweden)	grey metal
1925	Rhenium	W. Noddack, Fr I. Tacke, O. Berg (Germany)	white–grey metal
1937	Technetium	C. Perrier (France) E. Segre (Italy/USA)	silver–grey metal
1939	Francium	Mlle M. Percy (France)	metallic solid
1940	Astatine	D.R. Corson, K.R. Mackenzie (USA) E. Segre (Italy/USA)	metallic solid
1945	Promethium	J. Marinsky, L.E. Glendenin, C.O. Corgell (USA)	metallic solid

Figure 18.1 Mercury and bromine are liquid at room temperature.

Properties of elements and compounds

Only a very few of the substances you see around you are elements. The most common solid elements are metals such as aluminium and copper, though objects made of the elements gold and silver may be more obvious.

There are only two elements that are liquid at room temperature and standard pressure. They are mercury and bromine. Eleven elements are gases under normal conditions. Oxygen and nitrogen, which together form about 98% of the air, are two of them.

Each element has its own special properties. For example, sodium is a soft, silvery-white metal with a melting point of 97.86 °C and a boiling point of 884 °C and chlorine is a yellow–green gas with a melting point of −100.97 °C and a boiling point of −34.03 °C.

Most substances are made from two or more elements that are joined together. These substances are called compounds. They have properties which are different from the elements that make them. Common salt, for example, is a compound of sodium and chlorine and is a white solid with a melting point of 801 °C and a boiling point of 1420 °C.

Sorting out the elements

John Newlands (1838–1898) set out the elements in order of their atomic weights as found by experiment, starting with the lowest. (The atomic weight of an element is how many times the mass of one atom is greater than the mass of a hydrogen atom.) When Newlands looked at some of the elements that were eight spaces apart he discovered that they had similar properties. Moving down the list in this way he found that some of the properties reappeared periodically.

Dmitri Mendeleev (1834–1907) also noticed how the properties of the elements varied periodically and rearranged the elements into a table known as the periodic table. He found that elements in the columns had similar properties and he called these columns of elements groups. Mendeleev assumed that there were still elements to be discovered and so left gaps in the table where he thought they would eventually be placed. He could predict the properties of the missing elements from the arrangement of the elements in the table. Eventually the missing elements were discovered and were found to have the properties that Mendeleev predicted. Over the years the periodic table has been revised. Today the elements are arranged in order of atomic number (see below).

5 What did Newlands and Mendeleev see in the table of atomic weights?

6 How did the discoveries of elements made after Mendeleev had produced his table show him to be right?

The modern periodic table

Each element is made of atoms. At the centre of an atom is the nucleus, which is composed of two kinds of particles called protons and neutrons. Around the nucleus are particles called electrons.

In the nucleus of each atom of each element there is a certain number of protons. This number is different from the number of protons in the nuclei of any other element's atoms. The number of protons in an atom is called the atomic number. In the modern periodic table elements are arranged in order of their atomic number (Figure 18.2).

1 H hydrogen

I	II											III	IV	V	VI	VII	VIII
																	2 He helium
3 Li lithium	4 Be beryllium											5 B boron	6 C carbon	7 N nitrogen	8 O oxygen	9 F fluorine	10 Ne neon
11 Na sodium	12 Mg magnesium											13 Al aluminium	14 Si silicon	15 P phosphorus	16 S sulphur	17 Cl chlorine	18 Ar argon
19 K potassium	20 Ca calcium	21 Sc scandium	22 Ti titanium	23 V vanadium	24 Cr chromium	25 Mn manganese	26 Fe iron	27 Co cobalt	28 Ni nickel	29 Cu copper	30 Zn zinc	31 Ga gallium	32 Ge germanium	33 As arsenic	34 Se selenium	35 Br bromine	36 Kr krypton
37 Rb rubidium	38 Sr strontium	39 Y yttrium	40 Zr zirconium	41 Nb niobium	42 Mo molybdenum	43 Tc technetium	44 Ru ruthenium	45 Rh rhodium	46 Pd palladium	47 Ag silver	48 Cd cadmium	49 In indium	50 Sn tin	51 Sb antimony	52 Te tellurium	53 I iodine	54 Xe xenon
55 Cs caesium	56 Ba barium	57 La lanthanum	72 Hf hafnium	73 Ta tantalum	74 W tungsten	75 Re rhenium	76 Os osmium	77 Ir iridium	78 Pt platinum	79 Au gold	80 Hg mercury	81 Tl thallium	82 Pb lead	83 Bi bismuth	84 Po polonium	85 At astatine	86 Rn radon

metals   non-metals   noble gases

Figure 18.2 Part of the modern periodic table.

Many of the columns of elements in the periodic table are called groups. The elements in a group share similar properties. A trend can be seen in the properties as you go down the group.

Group I, the alkali metals

The metals in this group are not alkalis, but the oxides and hydroxides that they form are. It is this property of these compounds that gives the metals in this group their name. Table 18.2 shows some of the physical properties of the alkali metals.

Table 18.2 Physical properties of the alkali metals.

Element	Density (g/cm³)	Melting point (°C)	Boiling point (°C)
Lithium	0.53	180.6	1344
Sodium	0.97	97.9	884
Potassium	0.86	63.5	760
Rubidium	1.53	39.3	688
Caesium	1.90	28.5	671

A closer look at the alkali metals

Lithium

Lithium's name is derived from lithis, the Greek word for stone, because it is found in many kinds of igneous rock. It is used in batteries and in compounds used as medicines to treat mental disorders.

Sodium

Metallic sodium is used in certain kinds of street lamp that give an orange glow. It is alloyed with potassium to make a material for transferring heat in a nuclear reactor. Sodium compounds such as sodium hydroxide have a wide range of uses. In the body sodium is needed by nerve cells. They use it in the transfer of electrical signals called nerve impulses.

Potassium

Potassium is used to make the fertiliser potassium nitrate. In the body it is used for the control of the water content of the blood and is used with sodium in sending electrical signals by nerve cells.

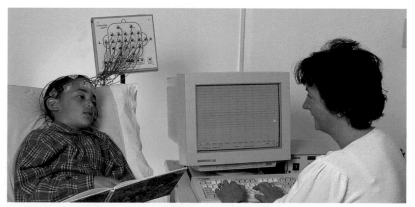

Figure 18.3 Measuring brain waves – nerve impulses are due to the movement of sodium and potassium ions in brain cells.

7 Which of these statements about the trends in Table 18.2 are true?
   a) The density of the metals generally
      i) increases,
      ii) decreases down the group.
   b) The melting point of the metals generally
      i) increases,
      ii) decreases down the group.
   c) The boiling point of the metals generally
      i) increases,
      ii) decreases down the group.
8 Which element does not follow a trend? Describe how it differs from the trend.
9 Sodium is a softer metal than lithium. Describe how you think the softness of potassium and rubidium compare with that of sodium.
10 Which metal has the smallest temperature range for its liquid form?
11 Look at the information about sodium and potassium in the reactivity series (Table 21.1, page 192) and predict a position for
   a) lithium and
   b) rubidium in the series.

Rubidium

Rubidium gets its name from the Latin word ruber, which means red. This describes the lines produced by rubidium when it is examined with a device called a spectroscope. Rubidium is used in the filaments of photoelectric cells which convert light energy into electrical energy.

Caesium

Caesium's name comes from the Latin word caesius, which means bluish-grey. This describes the colour of the lines the metal produces when examined with a spectroscope. Caesium is used in photoelectric cells and as a time keeper in atomic clocks. The vibration of the atoms is used to measure time very accurately. Each atom vibrates over nine thousand million times a second.

Group II, the alkaline earth metals

These metals are not alkalis but their oxides and hydroxides dissolve slightly in water to make alkaline solutions. Table 18.3 shows some of the physical properties of these metals.

Table 18.3 Properties of the alkaline earth metals.

Element	Density (g/cm^3)	Melting point (°C)	Boiling point (°C)
Beryllium	1.85	1289	2476
Magnesium	1.74	649	1097
Calcium	1.53	840	1493
Strontium	2.58	768	1387
Barium	3.60	729	1880

A closer look at the alkaline earth metals

Beryllium

Beryllium combines with aluminium, silicon and oxygen to make a mineral called beryl. Emerald and aquamarine (Figure 18.4) are two varieties of beryl which are used as gemstones in jewellery.

Beryllium is mixed with other metals to make alloys that are strong, yet light in weight. It is also used in a mechanism that controls the speed of neutron particles in a nuclear reactor.

Figure 18.4 Emerald (left) and aquamarine (right).

12 Which of these statements about the trends in Table 18.3 are true?
  a) The density of the metals generally
    i) increases,
    ii) decreases down the group.
  b) The melting point of the metals generally
    i) increases,
    ii) decreases down the group.
  c) The boiling point of the metals generally
    i) increases,
    ii) decreases down the group.

For discussion

How do the trends shown in Table 18.3 compare with those shown in Table 18.2?

Magnesium

Magnesium is used in fireworks to make a brilliant white light. Another important use is to mix it with other metals to make strong, lightweight alloys such as those used to make bicycle frames.

Green plants need magnesium in order to make the chlorophyll that traps the energy from sunlight in photosynthesis. Magnesium is needed in the body for the formation of healthy bones and teeth.

Calcium

Calcium's name is derived from the word calx, which is the Latin name for the substance lime. Lime is actually calcium oxide. Calcium forms many compounds with a wide range of uses, from baking powders and bleaching powders to medicines and plastics. In the human body calcium is required for the formation of healthy teeth and bones and for the contraction of muscles.

Strontium

Strontium forms salts which make a red flame when they burn. They are used in flares for signalling the position of survivors of shipwrecks or mountaineering accidents, and to make the red colour in fireworks. Strontium has radioactive isotopes which are produced in nuclear reactions.

Barium

Barium has a wide range of uses, from safety matches and providing the green colour in fireworks to mixing with other metals to make alloys. It is best known in a form called a barium meal (see Figure 18.5). This substance is barium sulphate and it forms a suspension that stops X-rays passing through it.

A barium meal is used in medicine to examine the alimentary canal of a patient. The patient eats the barium meal and as it passes along the alimentary canal the patient's body is X-rayed. The outline of the alimentary canal and the position of the barium meal can be seen in the X-ray photographs. The pictures help the doctors to diagnose the patient's condition.

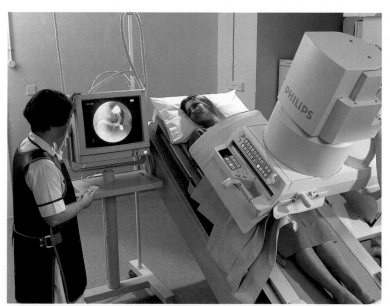

Figure 18.5 Examining an X-ray of a patient's abdomen using a barium meal.

13 What trends can you see in the melting points and the boiling points of the halogens?

14 Using the information in Table 18.4, deduce which halogens are
   a) solids,
   b) liquids or
   c) gases
   at room temperature.
   Explain your answers.

Group VII, the halogens

The word halogen is a Greek word for salt former and all the elements in this group form salts readily. Table 18.4 shows some of the properties of these elements.

Table 18.4 Physical properties of the halogens.

Element	Melting point (°C)	Boiling point (°C)
Fluorine	−219.7	−188.2
Chlorine	−100.9	−34.0
Bromine	−7.3	59.1
Iodine	113.6	185.3
Astatine	302	377

A closer look at the halogens

Fluorine

Fluorine is a pale yellow–green poisonous gas. It is found in combination with calcium in the mineral fluorite. This mineral glows weakly when ultraviolet light is shone on it. This property is called fluorescence. One variety of fluorite called Blue John has coloured bands and is carved into ornaments.

Fluorine is combined with hydrogen to make hydrogen fluoride, which dissolves glass and is used in etching glass surfaces. Sodium fluoride prevents tooth decay and is added to some drinking water supplies. Fluorine is one of the elements in chlorofluorocarbons or CFCs (see page 204).

Chlorine

Chlorine is a yellow–green poisonous gas. It is found in combination with sodium as rock salt. Chlorine is used to kill bacteria in water supply systems and is also used in the manufacture of bleach. It forms hydrochloric acid which has many uses in industry.

Bromine

Bromine is a red–brown liquid which produces a brown vapour at room temperature that has a strong smell and is poisonous. Bromine is extracted from bromide salts in sea water and is used, with silver, in photography. Silver bromide is light sensitive and is used in photographic film to record the amount of light in different parts of the image focused by the camera lens.

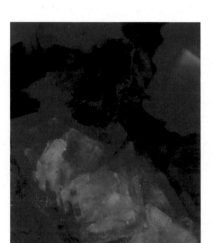

Figure 18.6 Fluorite glowing.

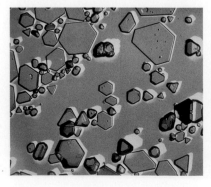 

Figure 18.7 Magnified images showing silver bromide crystals on a piece of photographic film (left) and silver deposits on a developed film (right).

15 Fluorine is more reactive than chlorine, chlorine is more reactive than bromine and bromine is more reactive than iodine. Is this trend shared by the alkali metals and alkaline earth metals? Explain your answer. (Look at the reactivity series, Table 21.1 on page 192, to help you answer.)

Iodine

Iodine is a grey–black solid. It is extracted from iodine salts in sea water and is used as an antiseptic and also in photography. Potassium iodide solution is used to detect starch in food tests. It is needed by the body for the production of a hormone which acts as a catalyst in oxidation reactions in the cells.

Astatine

This element is radioactive. It has many isotopes but they are all unstable and eventually break down into other elements.

Summary

- Substances can be broken down into elements (*see page 163*).
- Each element is composed of atoms (*see page 166*).
- An atom contains protons, neutrons and electrons (*see page 166*).
- The elements are arranged in the periodic table in order of their atomic number (number of protons) (*see page 166*).
- Group I of the periodic table contains the alkali metals (*see page 167*).
- Group II of the periodic table contains the alkaline earth metals (*see page 168*).
- Group VII of the periodic table contains the halogens (*see page 170*).

End of chapter question

1 Elements in the alkali metals, alkaline earth metals and halogens are important in our lives. How accurate is this statement? Explain your answer.

19 Compounds and mixtures

Mixing elements

Each element has its own particular properties. Sulphur, for example, is yellow and if shaken with water it will tend to float. Iron is black and magnetic and produces hydrogen when it is placed in hydrochloric acid. If the two elements are mixed together, a grey–black powder is produced. The colour depends on the amount of sulphur mixed with the iron. Although the two elements are close together, their properties do not change. If a magnet is stroked over the mixture, iron particles leap up and stick to it. If the mixture is shaken with water the sulphur will tend to float.

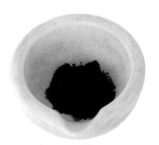

Figure 19.1 Black iron (left) and yellow sulphur (centre) mix to form a grey–black powder (right).

From elements to a compound

However, if the mixture of iron and sulphur is heated a change takes place. The atoms of iron and sulphur join together and form a compound called iron sulphide. It does not have the yellow colour of the sulphur or the magnetic properties of the iron. It has its own properties – it is a black non-magnetic solid. All compounds have properties which differ from the elements that formed them.

Figure 19.2 The formation of iron sulphide.

The reaction which takes place when iron and sulphur are heated is shown in the word equation:

iron + sulphur → iron sulphide

---

1 Add the state symbols to the equation for the reaction of iron and sulphur.

Looking closer at chemical reactions

Atoms of carbon and hydrogen join together to make compounds called hydrocarbons. The hydrocarbon methane is the main gas in natural gas. The word equation for the burning of methane is:

methane(g) + oxygen(g) → carbon dioxide(g) + water(l)

When you look at the equation, you can try and visualise the molecular structure of the reactants and products. If you do, you may think of something like Figure 19.3.

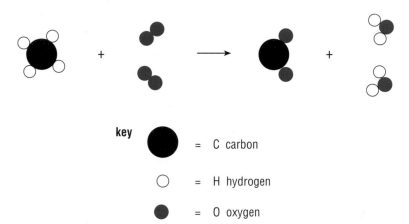

key   = C carbon

    = H hydrogen

    = O oxygen

Figure 19.3   The molecules involved in the burning of methane.

When you count up the numbers of different atoms in the reactants and products you find that they are the same. No atoms have been lost or gained. They have just been rearranged. There has been no creation of particles of matter, nor have any particles of matter been destroyed. Matter has been conserved. As each particle has mass this also means that in a chemical reaction mass is conserved.

Synthesis reactions

When two or more substances take part in a chemical reaction to make one compound, the reaction is called a synthesis. When a compound forms it may be in a different state to the elements from which it formed. For example, when water forms from hydrogen and oxygen the two gases produce a liquid. The word equation for this reaction is:

hydrogen(g) + oxygen(g) → water(l)

Elements from different states may take part in a chemical reaction to make a compound. The carbon in a block of glowing charcoal on a barbecue combines with oxygen in the air to make carbon dioxide. The word equation for this reaction is:

carbon(s) + oxygen(g) → carbon dioxide(g)

When magnesium is heated in air, the oxygen in the air combines with the magnesium to form magnesium oxide:

magnesium(s) + oxygen(g) → magnesium oxide(s)

---

2 When magnesium is placed in acid, it eventually disappears as bubbles are produced.
   a) Is mass conserved?
   b) Explain your answer.
   c) What experiment could you perform to support your answer?

3 Is mass conserved when magnesium burns in air? Explain your answer.

During the reaction of magnesium and oxygen large amounts of heat and light are released. This synthesis reaction is a major feature of all firework displays (see Figure 19.4).

If, on burning magnesium in air, you weighed the magnesium before and after it was burned, you would find that the mass increases.

Figure 19.4 A firework display.

Proportions

In a compound, the elements are always present in the same proportions. For example, in iron sulphide there is always one atom of iron joined with one atom of sulphur. Two atoms of iron do not sometimes join with one atom of sulphur, nor does one atom of iron join with three atoms of sulphur. The proportion of one element to the other is always the same. The elements in a compound are said to occur in fixed proportions.

A mixture may be made from elements (for example iron and sulphur) or compounds (for example iron sulphide and magnesium oxide) or even a mixture of the two (for example magnesium oxide and sulphur). Whatever the substances in a mixture, the proportions can vary widely – the proportions are not fixed.

Some chemical reactions of compounds

Decomposition reactions

One of the simplest types of reaction is called decomposition. There is only one reactant and it breaks down into two or more products.

Thermal decomposition

4 How is a decomposition reaction different from a synthesis reaction?

The most common type of decomposition is called thermal decomposition. In this case a compound is heated and it breaks down into other substances – the products. For example, when silver oxide is heated it breaks down into silver and oxygen.

Figure 19.5 The 'limelight man' in an old theatre.

Heating limestone

A simple thermal decomposition can be performed by heating a small piece of limestone. Limestone is made of a compound called calcium carbonate. The heat breaks down the compound into calcium oxide and carbon dioxide. The word equation for this reaction is:

calcium carbonate → calcium oxide + carbon dioxide

Calcium oxide does not break down when it is heated, but at very high temperatures it becomes incandescent and gives out a bright white light known as limelight. This was used to light the stages of theatres before electricity was available and gave rise to the expression 'in the limelight' (see Figure 19.5).

Heating copper carbonate

Copper carbonate is a green powder. If it is heated strongly it breaks down into black copper oxide and carbon dioxide. The word equation for this reaction is:

copper carbonate → copper oxide + carbon dioxide

5 Write the equation for the decomposition of calcium carbonate with state symbols.
6 Write the equation for the decomposition of copper carbonate with state symbols.

Figure 19.6 Before (left) and after (right) the heating of copper carbonate.

Heating potassium manganate(VII)

Potassium manganate(VII) is a purple crystalline solid which is a compound of potassium, manganese and oxygen. When it is heated it breaks down and releases oxygen. The gas can be collected as shown in Figure 19.7.

Heating some metal oxides

Not all compounds break down when they are heated. For example, when copper oxide, magnesium oxide and zinc oxide are heated each metal remains combined with oxygen.

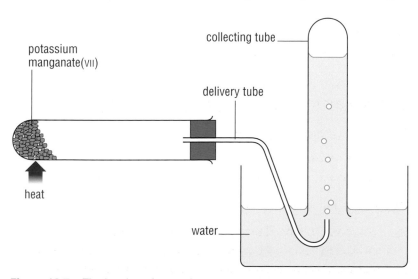

Figure 19.7 The heating of potassium manganate(VII).

Decomposition by light

When light shines on silver chloride it decomposes into tiny black crystals of silver metal and chlorine gas. The word equation for this reaction is:

silver chloride(s) → silver(s) + chlorine(g)

See page 170 for the use of this kind of reaction in photography.

Some compounds used in dyes to colour fabrics in clothes and curtains are also changed by the action of light (Figure 19.8).

Figure 19.8 The dye of the fabric under the collar, which is usually covered, has not been affected by sunlight.

Decomposition by electricity

Electricity can be used to decompose substances which have been melted and are in liquid form, or are in solution. Water (molten ice) can be decomposed into hydrogen and oxygen. The equation for this reaction is:

water(l) → hydrogen(g) + oxygen(g)

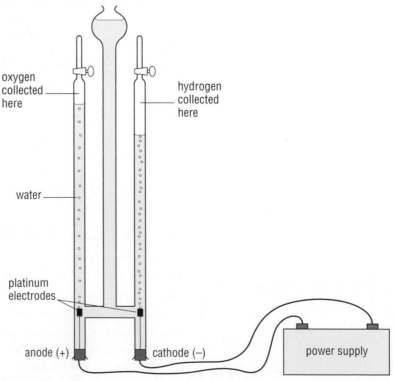

Figure 19.9 Apparatus for the decomposition of water.

Neutralisation

Neutralisation reactions were introduced on page 144. These reactions occur between acids and bases. Acids and bases are compounds. They have properties which are 'opposite' in nature to each other. For example, acids turn a substance called blue litmus from blue to red while bases turn red litmus to blue. Equal quantities of acid and base can be brought together to produce a neutralisation reaction in which the products have properties different from either of the reactants.

The word equation for the neutralisation reaction between hydrochloric acid and sodium hydroxide is:

hydrochloric + sodium → sodium + water
acid   hydroxide   chloride

7 Three of the compounds in the neutralisation equation are dissolved in water. Use this information to write the equation including the state symbols.

Precipitation

Some compounds dissolve in liquids to make solutions. When some solutions are mixed together a precipitate is formed. This is formed by tiny particles that do not dissolve in the mixture of the solutions. The precipitate makes the liquid cloudy.

If silver nitrate solution is poured into a solution of sodium chloride, a chemical reaction takes place which produces silver chloride. This forms a white precipitate.

silver   + sodium   → silver   + sodium
nitrate(aq)   chloride(aq)   chloride(s)   nitrate(aq)

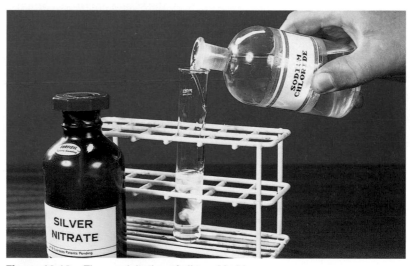

Figure 19.10   The precipitation of silver chloride.

Air – a mixture of gases

The air that we breathe is part of the atmosphere, which is a mixture of gases that covers the surface of the Earth. It stretches out into space for about 1000 km.

It is believed that the atmosphere was produced by a process known as out-gassing. In this process, gases from inside the Earth are released through volcanoes (see page 203). It began when the Earth formed and continues to the present day.

8 How many different kinds of mixtures can you think of? Check your answers with pages 153 and 154.

COMPOUNDS AND MIXTURES

The atmosphere is divided into five layers (see Figure 19.11). The composition of the gases in the atmosphere changes as you pass through the layers from outer space to the Earth's surface.

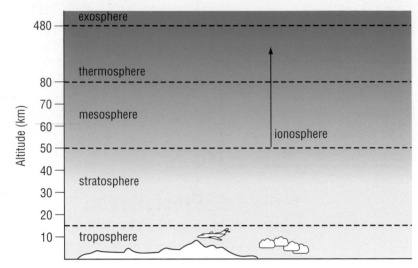

Figure 19.11 The layers of the atmosphere.

In the highest layer of the atmosphere, the exosphere, the mixture of gases is 25% helium and 75% hydrogen. As you sink through the thermosphere and mesosphere, the amount of hydrogen in the atmosphere falls to zero, the amount of helium falls to 15% and the amounts of nitrogen and oxygen rise to 70% and 15% respectively.

In the stratosphere the composition of gases is 1% ozone, 1% argon, 18% oxygen and 80% nitrogen. The composition of the air in the troposphere is shown in Figure 19.12.

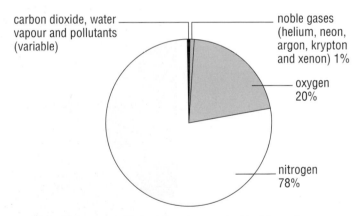

Figure 19.12 The composition of the air near the Earth's surface.

Uses of the air gases

All the gases that make up the air are colourless and do not have a smell, but they have many uses.

Nitrogen

Nitrogen hardly dissolves in water. It is a neutral gas and is very unreactive with other chemicals, although at the very high temperatures found in a lightning flash it combines with oxygen to form nitrogen dioxide.

Figure 19.13 Storing a sample in liquid nitrogen.

As nitrogen is so unreactive it can be used in its gaseous form to replace air in food packaging. If nitrogen is cooled to below −196 °C it changes into a liquid. In its liquid form nitrogen is used as a coolant. The low temperatures achieved are used for storing biological tissues such as blood and semen, embryos and organs such as kidneys (see Figure 19.13).

Nitrogen is an essential component of proteins, which are used to form the bodies of plants and animals. It is taken into plants through the roots in the form of minerals called nitrates, and passed on to animals when they feed on the plants. Nitrogen is used to make fertiliser to help crop plants grow.

Nitrogen is also used to make ammonia and nitric acid which are used in the chemical industry.

Oxygen

Oxygen is a neutral gas. A litre (1000 cm^3) of air contains 200 cm^3 of oxygen. Oxygen dissolves in water, but a litre of water may only hold up to 10 cm^3 of oxygen. However, this is enough to support a wide range of different kinds of aquatic life. Oxygen is essential for the process of respiration. In this process, plants and animals release energy which they use to keep themselves alive. In hospitals, additional oxygen is given to patients with respiratory diseases to help them breathe. It is mixed with other gases and stored in cylinders which allow divers to swim underwater and mountaineers to climb at high altitudes where the concentration of oxygen in the air is lower than at sea level.

Figure 19.14 A sub-aqua diver.

Oxygen reacts with many substances in the process of burning. When acetylene is burned in oxygen, temperatures as high as 3200 °C can be achieved. This is hot enough to cut through most metals or to weld metals together. Oxygen is also used to burn fuel in rocket engines in space craft.

Noble gases

The noble gases are very unreactive.

Argon

Argon is used in light bulbs. When electricity passes through the tungsten wire in the filament the metal gets hot. If oxygen were present it would react with the hot tungsten and the filament would quickly become so thin that it would break. Argon is used instead of air containing oxygen because it does not react with the tungsten and the filament lasts longer.

It is also used in making silicon and germanium crystals for the electronics industry.

Neon

This gas produces a red light when electricity flows through it and is used in lights for advertising displays.

Figure 19.15   Advertising displays in New York.

Helium

Helium is lighter than air and is used to lift meteorological balloons into the atmosphere. These balloons carry equipment for collecting information for weather forecasting and relay it by radio to weather stations. Helium is also mixed with oxygen to help deep sea divers breathe underwater.

Krypton

This is used in lamps which produce light of a high intensity, such as those used for airport landing lights and in lighthouses.

Xenon

Xenon is used to make the bright light in a photographer's flash gun.

Figure 19.16   Launching a meteorological balloon.

9 'Air is a mixture of useful chemicals.' Is this description correct? Explain your answer.

Identifying substances

Every pure substance has a combination of melting point and boiling point that is different from those of other substances. These can then be used like a 'fingerprint' to identify the substance. Table 19.1 shows the melting and boiling points of some common substances.

Table 19.1 Melting points and boiling points.

Substance	Melting point (°C)	Boiling point (°C)
Nitrogen	−214	−196
Ammonia	−78	−33
Bromine	−7	59.1
Mercury	−39	357
Sodium chloride	801	1420
Iron	1539	2887

10 Which substances are gases at:
   a) 0 °C,
   b) 120 °C?
11 Which substance is still solid at 1000 °C?

Summary

- The elements in a mixture retain their properties (*see page 172*).
- When elements form a compound, they form a substance which has different properties from their own (*see page 172*).
- Mass in conserved in a chemical reaction (*see page 173*).
- A synthesis reaction occurs when two or more substances take part in a chemical reaction to make one compound (*see page 173*).
- The elements in a compound are always present in the same proportions (*see page 174*).
- Compounds can be decomposed by heat (*see page 174*).
- Some compounds can be decomposed by light (*see page 176*).
- Water and many other substances can be decomposed by electricity (*see page 176*).
- A neutralisation reaction occurs when equal quantities of acid and base are brought together (*see page 177*).
- A precipitate is formed when some solutions are mixed together (*see page 177*).
- The atmosphere is divided into five layers (*see page 178*).
- Nitrogen, oxygen and the noble gases in the air have many uses (*see page 178*).

End of chapter questions

1 When magnesium is heated in air it forms a compound.
   a) What is this compound?
   b) Is the compound heavier or lighter in weight than the original piece of magnesium?
   Explain your answers.

2 'Limestone decomposes into its elements when it is heated in air.'
   Is this statement true? Explain your answer.

3 When something burns in air, is all the air used up? Explain your answer.

20 Metals and metal compounds

Elements can be divided into two large groups, according to their properties. These groups are called metals and non-metals.

Comparing metals and non-metals

Physical properties

Table 20.1 shows the physical properties of the elements in the two groups.

Table 20.1 The physical properties of metals and non-metals.

Property	Metal	Non-metal
state at room temperature	solid (except mercury)	solid, liquid or gas
density	generally high	low
surface	shiny	dull
melting point	generally high	generally low
boiling point	generally high	generally low
effect of hammering	shaped without breaking	breaks easily
effect of tapping	a ringing sound	no ringing sound
strength	high	generally very weak
magnetic	a few examples	no examples
conduction of heat	good	poor
conduction of electricity	good	poor

A material through which electricity can pass is called an electrical conductor. A material through which electricity cannot pass is called an electrical insulator.

A material through which heat can pass is called a conductor of heat. A material through which heat cannot pass is called an insulator.

Non-metals have a wider range of physical properties than metals because nearly all metals are solids at room temperature and non-metals can either be solids, liquids or gases.

Using physical properties to group elements can be unreliable as a few elements have exceptional properties. Mercury is the only metal that is a liquid at normal room temperature, and iodine is a solid with a shiny surface that looks metallic even though it is a non-metal.

Carbon is an element that can exist in different crystalline forms, graphite and diamond. Diamond has a very high melting point and boiling point, while graphite conducts electricity.

Metals and non-metals can be more clearly identified by their chemical properties.

> **1** Why may physical properties be unreliable for grouping substances into metals and non-metals?

Figure 20.1 Diamond (left) and graphite (right).

Chemical properties

Some metals and non-metals react together to produce salts. These reactions are examples of synthesis reactions. For example, if a burning piece of sodium is placed in a jar of chlorine gas in a fume cupboard the two elements combine to make a white solid. The word equation for this reaction is:

sodium(s) + chlorine(g) → sodium chloride(s)

If zinc or copper is heated with sulphur the metal sulphide is formed. The word equations for these reactions are:

zinc(s) + sulphur(s) → zinc sulphide(s)

copper(s) + sulphur(s) → copper sulphide(s)

Oxygen is a non-metal and reacts with many metals and non-metals to form oxides.

Reaction with oxygen

If a metal takes part in a chemical reaction with oxygen, a metal oxide is formed. A metal oxide is a base, and forms a salt and water when it takes part in a chemical reaction with an acid. A few metal oxides are soluble in water. They are called alkalis. Calcium oxide is a soluble base (an alkali). This is the reaction that occurs between calcium oxide and water:

calcium oxide(s) + water(l) → calcium hydroxide(aq)

If a non-metal takes part in a chemical reaction with oxygen it also forms an oxide. Most oxides of non-metals are soluble. When they dissolve in water they form acids. Sulphur is a non-metallic element with a yellow crystalline form. If it is heated in air it burns and combines with oxygen to form sulphur dioxide, which is soluble in water. This reaction occurs between sulphur dioxide and water:

sulphur dioxide(g) + water(l) → sulphurous acid(aq)

When carbon powder is heated in air it glows red. If it is plunged into a gas jar of oxygen it becomes bright red. Carbon combines with oxygen to form carbon dioxide, which dissolves in water to form an acidic solution with a pH of 5.

Magnesium ribbon easily catches fire if it is held in a Bunsen burner flame, and burns with a brilliant white light if plunged into a gas jar of oxygen. Magnesium oxide (a white powder) is produced, which dissolves in water to make an alkaline solution with a pH of 8.

2 How may the reaction with oxygen be used to distinguish a metal from a non-metal?

3 Use the information in this section to decide whether
a) carbon,
b) magnesium
is a metal or a non-metal. Explain your answer.

Reaction of metals with acids

Figure 20.2 shows the reaction of a metal with an acid. There is zinc and hydrochloric acid in the flask. Bubbles of gas are emerging from the surface of the acid. The gas passes along the delivery tube and into the boiling tube. Here the gas pushes the water out of the way and in time may fill the tube.

4 How could you test the gas in the test-tube in Figure 20.2 to see if it was hydrogen?

Figure 20.2   Apparatus for the collection of hydrogen.

The word equation for this reaction is:

zinc(s) + hydrochloric acid(aq) → zinc chloride(aq) + hydrogen(g)

While it is easy to see the hydrogen gas as it forms bubbles and pushes water out of the way in the boiling tube, the zinc chloride cannot be seen. The reason for this is that zinc chloride is soluble. A solution of zinc chloride is forming in the flask. When the reaction is complete, the solution can be warmed so that evaporation takes place. The water changes to water vapour and escapes into the air and the zinc chloride is left behind.

Zinc chloride is a salt. Zinc also forms salts with sulphuric and nitric acids. The word equation for the reaction between zinc and sulphuric acid is:

zinc(s) + sulphuric acid(aq) → zinc sulphate(aq) + hydrogen(g)

Figure 20.3   Zinc chloride.

5 What is the word equation for the reaction between zinc and nitric acid? Write it with state symbols.
6 Other metals such as magnesium, aluminium and iron react with acid in a similar way to zinc. Write a general word equation which describes the reaction between a metal and an acid.

For discussion

There is very little hydrogen produced when sulphuric acid is added to calcium. The salt that is produced, calcium sulphate, is insoluble. Use this information to explain why the reaction does not take place for long.

7 Write the word equation for the reaction between copper carbonate and nitric acid with state symbols.

8 Write the word equation for the reaction between sodium carbonate and hydrochloric acid.

9 Write a general word equation which describes the reaction between a metal carbonate and an acid.

10 Write the word equation for the reaction between zinc oxide and hydrochloric acid.

11 Write a general word equation which describes the reaction between a metal oxide and an acid.

Reaction of acids with metal carbonates

The apparatus shown in Figure 20.2 can also be used to investigate the reaction between a metal carbonate and an acid. However, the gas that is produced in this reaction is not hydrogen. It is carbon dioxide.

When nitric acid is added to copper carbonate the following word equation describes the reaction:

copper   +   nitric   →   copper + water + carbon
carbonate   acid   nitrate   dioxide

Reactions of acids with metal oxides

Copper oxide is a black powder but when sulphuric acid is added to it a blue solution forms. This indicates that a chemical reaction has taken place. The word equation for this reaction is:

copper   +   sulphuric   →   copper   + water(l)
oxide(s)   acid(aq)   sulphate(aq)

If there is too much copper oxide present some will remain at the bottom of the beaker. The unreacted copper oxide can be removed by filtration. The copper sulphate can be removed from the solution as crystals, by allowing the water in the solution to evaporate.

Figure 20.4 Copper oxide (left), sulphuric acid (centre), and copper sulphate (right).

Figure 20.5 Copper sulphate crystals.

Chemical reactions and energy

When a chemical reaction takes place, energy may be given out or taken in.

Examples of reactions that give out energy are the burning of gas in a Bunsen burner and the explosion of petrol vapour in a car engine. Other chemical reactions, such as the reaction between acids and carbonates, give out heat too.

An example of a reaction that takes in energy is photosynthesis. This takes place in the leaves and other green parts of plants. Light energy is taken in from sunlight and is used as carbon dioxide and water combine to form sugar. The decomposition of limestone to make lime (see page 175) is another reaction that takes in energy. Heat is required to start and maintain the reaction.

Figure 20.6 Photosynthesis is a reaction that takes in energy from sunlight.

Some chemical reactions need a little energy to make them start and then they can continue on their own. The energy required to start the reaction is usually provided in the form of heat. The reaction taking place on a burning match head provides the energy to start the wax burning on a candle wick.

Neutralisation and salts

Neutralisation is a reaction which takes place between an acid and an alkali. The point at which neutralisation occurs can be found by using universal indicator solution or a pH probe and by carefully measuring the acid added to a certain volume of alkali.

Salts are prepared by neutralisation followed by evaporation of the neutralised solution, leaving salt crystals.

Some people have difficulty thinking about the word salt. All they can think about is sodium chloride – common salt. There are many different salts and some of them are useful. For example, potassium nitrate is

12 What changes in energy can occur when chemical reactions take place?

13 Why is it that you must put a match flame to a candle wick to light it but can remove the match when the wick starts to burn?

14 Write the word equation for the reaction taking place in the neutralisation of hydrochloric acid and sodium hydroxide.

METALS AND METAL COMPOUNDS

186

used as a fertiliser, as a preservative of meat products and in making gunpowder. Magnesium sulphate is used as a laxative to ease constipation. Sodium stearate is a salt used to make soap.

Passage of electricity through metals and non-metals

A solid can be tested to find out if it conducts electricity by using a circuit like the one shown in Figure 20.7. The solid to be tested is secured between the pair of crocodile clips and the switch is closed. The lamp lights if the solid conducts electricity. By using the circuit, metals and the non-metal carbon, in the form of graphite, are found to conduct electricity. Other non-metals such as sulphur and solid compounds such as sodium chloride do not.

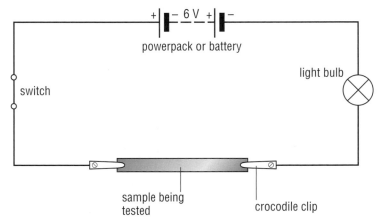

Figure 20.7   A circuit for testing conduction of solid materials.

If a liquid is to be tested, the apparatus shown in Figure 20.8 is used. Graphite (carbon) rods are attached to the crocodile clips. The ends of the rods are then lowered into the liquid. The liquid may be a pure substance or a solution.

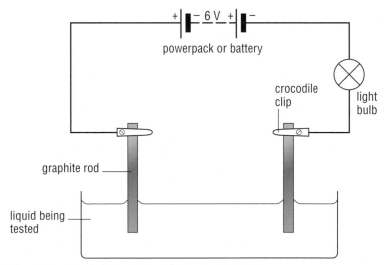

Figure 20.8   A circuit for testing conduction of liquids.

Summary

- Metals have different properties from non-metals (*see page 182*).
- Some metals and non-metals react together to produce salts (*see page 183*).
- Some metals and non-metals react with oxygen to produce oxides (*see page 183*).
- Metals react with acids (*see page 184*).
- Metal carbonates react with acids (*see page 185*).
- Metal oxides react with acids (*see page 185*).
- The point of neutralisation between an acid and an alkali can be found by using an indicator (*see page 186*).
- Salts have many uses (*see page 186*).
- Metals and the non-metal carbon, but not their compounds, conduct electricity (*see page 187*).

End of chapter questions

1 You are given a Bunsen burner, a hammer, a magnet and a battery with a lamp, wires and switch.

   a) How could you use them to distinguish between samples of metals and non-metals?

   b) Assess the usefulness of the tests you make.

2 Write word equations with state symbols to describe how

   a) sodium reacts with chlorine,

   b) copper reacts with sulphur,

   c) aluminium reacts with hydrochloric acid,

   d) zinc carbonate reacts with sulphuric acid,

   e) magnesium oxide reacts with nitric acid.

21 *Patterns of reactivity*

All metals do not react in the same way with oxygen, water and acids. The way they react forms a pattern.

Reaction of metals with gases in the air

Silver reacts with sulphur in the air to form a coating of silver sulphide, which is black, on the surface. This reduces the reflectivity of the metal and the silver is said to be tarnished.

Copper reacts with substances in the air and slowly loses its shiny brown surface. The surface of copper used on roofs forms a substance called verdigris. This contains copper carbonate.

Sodium is a soft metal and can be cut with a knife. If the freshly cut surface is left exposed to the air it tarnishes quickly.

Rust

When water vapour from the air condenses on iron or steel it forms a film on the surface of the metal. Oxygen dissolves in the water and reacts with the metal to form iron oxide. This forms brown flakes of rust which break off from the surface and expose more metal to the oxygen dissolved in the water. The iron or steel continues to produce rust until it has completely corroded.

Steel is used for making girders that support bridges and for making many parts of cars. If the steel is not protected it soon begins to rust. This weakens the metal. It makes bridges unsafe. It makes holes in car bodies and weakens the joints that hold the cars together, making them unsafe for use.

Figure 21.1 A rusted gate post.

1 Many tall buildings have a framework made of steel girders on which walls of brick and glass are built. If the steel was unprotected what would you expect to happen in time? Explain your answer.

Rust prevention

Rust can be prevented by keeping oxygen away from the iron or steel surface. This can be done by painting the surface or covering it in oil. However, if the paint becomes chipped or the oil is allowed to dry up, rust can begin to form. Steel can also be protected by covering the surface with chromium in a process called chromium plating. Steel used for canning foods is coated in a thin layer of tin.

The steel used for girders to build office blocks and bridges is coated in zinc in a process called galvanising or zinc plating.

Reaction with water

Here are some descriptions of the reactions that take place between water and metals. In the study of these reactions the metals were first tested with cold water. If there was no reaction, the test was repeated with steam (see Figure 21.2, overleaf).

Calcium sinks in cold water and bubbles of hydrogen form on its surface, slowly at first. The bubbles then increase in number quickly and the water becomes cloudy as calcium hydroxide forms. The bubbles of gas can be collected by placing a test-tube filled with water over the fizzing metal. The gas pushes the water out of the test-tube. If the tube, now filled with gas, is quickly raised out of the water and a lighted splint

2 Arrange the metals in this section in order of their reactivity with water.

3 Which metals would not be put into the apparatus in Figure 21.2 to see if they reacted with steam?

4 Which metals are less dense than water?

5 Water is a compound of hydrogen and oxygen which could be called hydrogen oxide. When hydrogen is released as a metal reacts with steam, what do you think is the other product of the reaction?

6 In the home, copper is used for the hot water tank and steel (a modified form of iron – see page 197) is used to make the cold water tank. Why can steel not be used to make the hot water tank?

held beneath its mouth, a popping sound is heard. The hydrogen in the tube combines with oxygen in the air and this explosive reaction makes the popping sound.

Copper sinks in cold water and does not react with it. Neither does it react with steam.

Sodium floats on the surface of water and a fizzing sound is heard as bubbles of hydrogen gas are quickly produced around it. The production of the gas may push the metal across the water surface and against the side of the container, where the metal bursts into flame. A clear solution of sodium hydroxide forms.

Iron sinks in water and no bubbles of hydrogen form. When the metal is heated in steam, hydrogen is produced slowly.

Magnesium sinks in water. Bubbles of hydrogen are produced only very slowly and a solution of magnesium hydroxide is formed. When the metal is heated in steam hydrogen is produced quickly.

Potassium floats on water and bursts into flames immediately. Hydrogen bubbles are rapidly produced around the metal. A clear solution of potassium hydroxide forms.

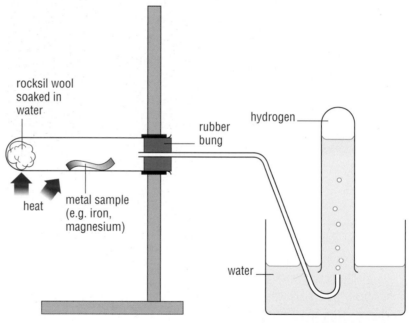

Figure 21.2   The apparatus to test the action of steam on a metal.

Reaction with acids

Here are some descriptions of the reactions that take place between different metals and hydrochloric acid. The metals were first tested with dilute hydrochloric acid.

Lead did not react with dilute hydrochloric acid.

Zinc reacted quite slowly with dilute hydrochloric acid to produce bubbles of hydrogen and a solution of zinc chloride was formed.

Figure 21.3   The reaction of magnesium with dilute hydrochloric acid produces hydrogen bubbles.

7 Arrange the metals in this section in order of their reactivity with hydrochloric acid.

8 Write a general word equation for the reaction between a metal and hydrochloric acid.

9 Arrange the metals mentioned in this section in order of their reactivity with oxygen. Start with what you consider to be the most reactive metal.

Copper did not react with dilute hydrochloric acid.

Magnesium reacted quickly with dilute hydrochloric acid and formed bubbles of hydrogen (see Figure 21.3) and a solution of magnesium chloride.

Iron reacted slowly with dilute hydrochloric acid to produce bubbles of hydrogen and a solution of iron chloride was formed.

Reaction with oxygen

Here are some descriptions of the reactions that take place when certain metals are heated with oxygen. When a substance combines with oxygen to form an oxide, this is called oxidation.

Copper develops a covering of a black powder without glowing or bursting into flame. Iron glows and produces yellow sparks; a black powder is left behind. Sodium only needs a little heat to make it burst into yellow flames and burn quickly to leave a yellow powder behind (see Figure 21.4). Gold is unchanged.

Figure 21.4 Sodium burning in a gas jar of oxygen (left). Sodium oxide powder is left behind (right).

Figure 21.5 Copper wire coils in silver sulphate solution. Silver is formed on the wire.

Displacement reactions

When metals react with acids, they displace hydrogen from the acid and form a salt solution. In a similar way, a more reactive metal can displace a less reactive metal from a salt solution of the metal.

When a copper wire is suspended in a solution of silver sulphate, the copper dissolves into the solution to form copper sulphate and silver metal comes out of the solution and settles on the wire (see Figure 21.5).

10 From the information about these two displacement reactions, arrange copper, silver and iron in order of reactivity – starting with the most reactive.

Look at Table 21.1 below to answer these questions.

11 How do you think the reactions that zinc makes with oxygen, water and acid compare with those that iron makes?

12 Would you expect zinc to displace
   a) iron,
   b) lead,
   c) aluminium in displacement reactions?
   Explain your answers.

If an iron nail is placed in copper sulphate solution, the iron dissolves to form a pale green iron sulphate solution and the copper comes out of the solution and coats the nail (see Figure 21.6).

Figure 21.6   This iron nail has been left in copper sulphate solution. Copper has formed on the nail.

The reactivity series

The reactivity series is a list of metals arranged in order of their reactivity, starting with the most reactive. The series is produced by studying the reactions of metals with oxygen, water, hydrochloric acid and solutions of metal salts. Table 21.1 shows 12 metals in the reactivity series and summarises their reactions with oxygen, water and hydrochloric acid.

Table 21.1   The reactivity series.

Metal	Reaction with oxygen	Reaction with water	Reaction with acid
potassium	oxide forms very vigorously	produces hydrogen with cold water	violent reaction – *not to be done in a school lab*
sodium			
calcium		produces hydrogen with steam	rate of reaction decreases down the table
magnesium			
aluminium			
zinc			
iron	oxide forms slowly		
tin	oxide forms without burning	no reaction with water or steam	very slow reaction
lead			
copper			no reaction
silver	no reaction		
gold			

Extracting and using metals

Metals differ from each other in the way they react with other elements and compounds. Some metals, such as gold and silver, are very unreactive and can be found in their metallic form in the Earth's crust. Elements which are found on their own in this way are called native elements. More reactive metals are found combined with other elements. A rock which is rich in a metal compound is called an ore.

Gold crystals   Silver crystals   Iron ore (haematite)   Aluminium ore (bauxite)

Figure 21.7   A selection of native metals and ores.

Different methods of extracting metals are used. They depend on the reactivity of the metal. For example, copper is quite unreactive and so can be extracted from its ore by roasting the ore in a furnace. Iron, which is more reactive, must be heated strongly in a blast furnace. This provides more heat energy which is needed for the reactions to release iron from its ore. Aluminium is more reactive than iron and needs even more energy to extract it. This energy is supplied by electricity.

In the sections that follow, the metals are arranged in order of their reactivity. The order starts with the least reactive (gold) and ends with aluminium – the most reactive of the metals that we use in large amounts.

Note on alloys

In the following descriptions of metals the word alloy will appear. Most metals in their pure form tend to be weak and soft. They are strengthened by mixing them with one or more other elements, usually different metals. These mixtures are called alloys. They are made by melting the metals, mixing them together and then allowing them to cool. The properties of an alloy can be changed by altering the proportions of the metals that are mixed together.

Gold

Gold is found on its own as a metal. Many pieces can occur together in veins (cracks in rocks) or they may be more spread out in the ores of other metals. When the rock bearing the gold is weathered, it is carried away in streams and rivers and the gold settles out with the rock particles where they are deposited.

Extraction

On a small scale, gold is extracted by panning (see Figure 21.8, overleaf). A sediment from a stream is put in a shallow, wok-shaped pan with some water and the pan is swirled. The water lifts the lighter rocks to the edge of the pan, where they escape, and the heavier particles of gold collect at the bottom.

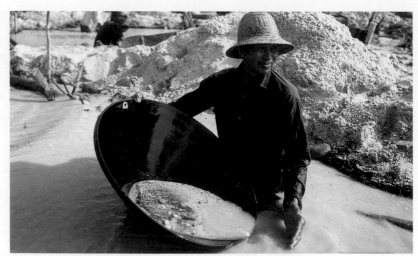

Figure 21.8 Panning for gold.

On a large scale, the deposits are mined. The ore which contains the gold is crushed and made into a powder. It is treated with potassium cyanide, in which the gold dissolves. The solution is filtered from rocky parts of the ore and the gold is removed from the solution by a displacement reaction.

Properties and uses

Gold does not react with the air and never loses its shine. It is soft and can be easily shaped or made into a thin sheet that does not break. The colour and shininess of gold make it an attractive material for jewellery. A very thin sheet of gold is called gold leaf and it is used for decoration of surfaces on buildings and books.

Gold is also used to make contacts that form the connections between wires in electrical circuits. This is because it does not corrode and this prevents the circuits from breaking.

The purity of gold is measured in carats. Pure gold is 24 carat gold and 18 carat gold is 75% pure gold.

Gold is alloyed with other metals. White gold is an alloy of gold, nickel and palladium. Rolled gold is a thin layer of a gold alloy that is bonded onto brass or nickel silver.

Copper

The ore from which copper is extracted is called chalcopyrite or copper pyrites. It contains copper, iron and sulphur, is brass yellow, and is found in igneous and metamorphic rocks.

Extraction

The ore is concentrated in a flotation cell (see Figure 17.8, page 157) and the copper, iron and sulphur are separated by roasting the ore in a furnace. The copper that is removed from the furnace still contains impurities. They are removed by making the copper into large slabs and hanging them in an electrical cell (see Figure 21.9). Each slab is an anode. As the electricity passes through the cell, the copper at the anode dissolves in the electrolyte and comes out of solution again on the cathode where it forms pure copper (see page 200). Gold and silver impurities in the metal fall to the bottom of the cell below the anode and form the anode sludge. They are removed and separated.

---

13 What is the purity of
  a) 22,
  b) 14,
  c) 9 carat gold?
14 Why do you think gold is alloyed with less expensive metals?

---

15 What are the three processes used in the extraction of copper?
16 What other metals are there in copper ore?

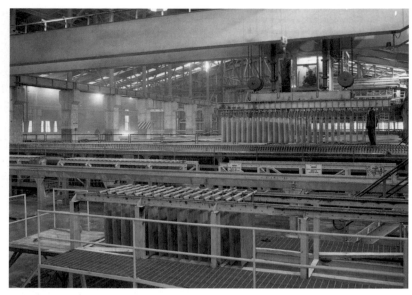

Figure 21.9 Making pure copper by electrolysis.

Properties and uses

Copper is a soft metal that can be easily shaped. It does not react with water so it is used to make water pipes, though large water pipes are often made of plastic which is cheaper. It conducts heat well and is used in the bases of some kinds of kitchen pan. Copper's softness also allows it to be pulled out into a wire and as it also conducts electricity well and corrodes very slowly, the wire can be used to conduct electricity inside buildings.

Copper is alloyed with tin to make bronze. This alloy was first made and used 5000 years ago and its name is used to describe a period of history in which a great many bronze implements were used – the Bronze Age. Bronze is a particularly sonorous metal (it makes a ringing sound) and it is used to make bells and cymbals because of the clear ringing sound it produces when it is struck.

Brass is an alloy of copper and zinc. It is strong, corrosion-resistant and is used to make the pins in electrical plugs. It is also a shiny metal and is used to make ornaments.

Iron

Iron combines with oxygen to form iron oxide. In the Earth's crust this compound frequently forms the mineral haematite. Sedimentary rocks which contain large amounts of haematite are important iron ores.

Extraction

Iron is separated from the oxygen in iron oxide by a reduction process. Reduction is the removal of oxygen from a compound by another substance.

The reduction of iron oxide takes place in a blast furnace (see Figure 21.11, overleaf). The iron ore is mixed with coke and limestone and tipped into the top of the blast furnace. Hot air is blown into the blast furnace through pipes close to the base of the furnace. The hot air causes the coke to ignite and burn strongly, raising the temperature as high as 2000 °C. The carbon in the burning coke reacts with oxygen in the hot air to form carbon dioxide. This gas rises through the hot coke

---

17 What properties of copper make it useful for electrical wiring in a home?

18 Why is brass better for use in plug pins than copper?

Figure 21.10 A 16th century Benin bronze.

Aluminium

The ore from which aluminium is extracted is called bauxite (see Figure 21.7, page 193). This rock is formed from a mixture of minerals that have been weathered in tropical regions of the world.

Extraction

Bauxite is dug up from the surface of the Earth's crust and is broken into small pieces in a crushing machine. The pieces are mixed with sodium hydroxide solution and the mixture is heated under pressure in a large sealed tank. The aluminium oxide from the ore dissolves in the solution to form sodium aluminate. The solution is filtered to remove the other rocky substances and allowed to cool.

In the cooling process, crystals of aluminium oxide form which are separated from the solution of sodium hydroxide. The crystals are then heated and the water of crystallisation escapes from them – leaving aluminium oxide in powdered form.

Aluminium is extracted from aluminium oxide by electrolysis. This means that the aluminium oxide must be in liquid form for the elements to be separated. The melting point of aluminium oxide is high and a great deal of energy would be needed to melt it, so an alternative method that uses less energy is used. The aluminium oxide is dissolved in molten cryolite – a mineral formed from sodium aluminium fluoride. The electricity is passed through this mixture in a cell as shown in Figure 21.14.

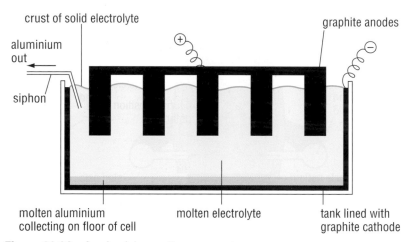

Figure 21.14   An aluminium cell.

Before the development of electrolysis for the extraction of aluminium, it was so difficult to extract that it was prized more than gold and silver and was the most expensive metal. Today, it is widely used because it is cheap to produce using electricity from hydroelectric power stations.

Properties and uses

Aluminium is soft, weak, light in weight and non-toxic. These properties make it useful for wrapping foods to keep them fresh. Aluminium is also a good conductor of electricity and, as it is light, these two properties make it useful for overhead power cables. Aluminium is also a good

27 Why is aluminium's lightness a particularly useful property?

28 As the aluminium in the aluminium oxide collects at the cathode, what would you expect to collect at the anode?

conductor of heat and, being lightweight, it is useful for making kitchen pans. The strength of aluminium is increased by mixing it with other metals. For example, aluminium is alloyed with copper to form a strong lightweight material for making aircraft and truck bodies and the frames for racing bicycles.

Aluminium is alloyed with copper and tin to make aluminium bronze. This is a strong, lightweight and corrosion-resistant metal which is used for fittings on the decks of boats and ships.

Discovery of the reactive metals

As we have seen, metals at the top of the reactivity series do not occur naturally on their own. They are always combined with other elements to form compounds. Very reactive metals are very difficult to separate from the elements they are combined with in a compound.

When people began to study chemicals scientifically, some compounds were thought to contain unknown elements but the experiments that were used at the time could not split up the compounds. With the invention by Volta in the 18th century of a device which could give a steady current of electricity (a battery), the process of electrolysis was developed and the electrical energy was found to be strong enough to break up the compounds and reveal new metals (see Davy's discoveries in Table 18.1 on page 164).

29 How do you think that reducing the melting point of an electrolyte makes the metal cheaper to produce?

In the extraction of sodium, for example, the electrolyte is molten sodium chloride. Sodium chloride has a high melting point and requires a large amount of energy to turn it from a solid into a liquid. Calcium chloride is added to the sodium chloride to lower its melting point and reduce the amount of energy required to melt it.

The molten electrolyte is passed into a Down's cell (see Figure 21.15) where sodium is collected from around the steel cathode and chlorine is collected from around the carbon anode.

Electrolysis is used on a large scale in the extraction of aluminium.

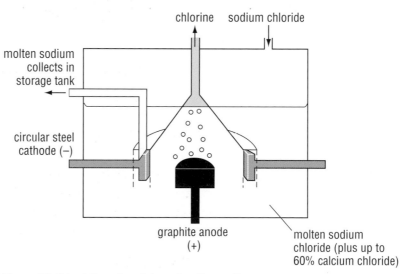

Figure 21.15 A Down's cell for extracting sodium.

Electroplating

Plating copper on copper

Two copper electrodes were weighed then dipped into a solution of copper sulphate and the current switched on for half an hour. No bubbles of gas were seen on either electrode and there appeared to be no other changes taking place. At the end of half an hour the current was switched off and the electrodes were cleaned and dried. When the electrodes were re-weighed the cathode was found to have gained weight and the anode was found to have lost weight. The weight lost by the anode was equal to the weight gained by the cathode. This 'plating' of copper on copper is used in the purifying of the metal after it has been extracted from its ore (see page 194).

Plating one metal with another

One metal can be coated with another by setting up electrodes and an electrolyte as shown in Figure 21.16.

The metal to be coated is the cathode and the metal to form the coating is made into the anode and is also present in the electrolyte. When the current of electricity is switched on, the metal from the anode dissolves in the electrolyte and metal in the electrolyte comes out of solution and forms a coating on the cathode.

This process is used to give objects made of a cheap metal a coating of a more expensive metal, to make them look more attractive. For example, it is used to coat cheap metals with gold to make jewellery or with silver to make EPNS (electroplated nickel silver) cutlery and ornaments.

Electroplating is also used to cover steel with chromium. The chromium gives the steel an attractive shiny surface and also protects the steel from rusting (see page 197).

> **30** What has happened at the cathode?
>
> **31** What has happened at the anode?

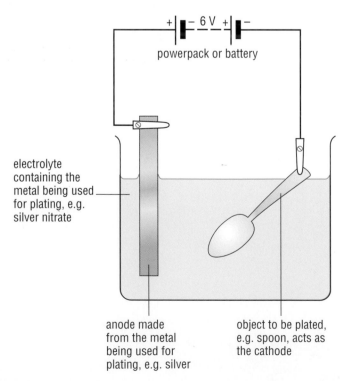

Figure 21.16 Electroplating.

Summary

- Iron forms rust with oxygen and water (*see page 189*).
- Some metals react vigorously with cold water while others do not react even with steam (*see page 189*).
- Some metals react with dilute acids (*see page 190*).
- Some metals react vigorously with oxygen while others do not react with it when heated in air (*see page 191*).
- Adding oxygen to a substance to form an oxide is called oxidation (*see page 191*).
- A more reactive metal can displace a less reactive metal from a salt solution of the less reactive metal (*see page 191*).
- Metals can be arranged in order of their reactiveness in the reactivity series (*see page 192*).
- The extraction and uses of metals depend on their properties (*see page 192*).
- Removal of oxygen from a compound by another substance is called reduction (*see page 195*).
- Electrolysis can be used to extract some reactive metals (*see pages 198 and 199*).
- Electroplating is the coating of one metal with another using electrolysis (*see page 200*).

End of chapter questions

1 Use Table 21.1 (page 192) to suggest the possible identities of these metals:

A – does not react with oxygen, water or acids.

B – produces hydrogen with cold water.

C – forms an oxide without burning but does not react with acids.

D – produces hydrogen with steam but only slowly forms an oxide with oxygen.

2 Construct a table that summarises the information about metals in this chapter. Use headings such as sources, extraction, and properties.

22 Chemistry and the environment

Humans in the environment

The first people used natural materials such as stone, wood, animal skins, bones, antlers and shells. They shaped materials using flint knives and axes. When they discovered fire they also discovered the changes that heat could make.

First they saw how it changed food and later it is believed that they saw how metal was produced from hot rocks around a camp fire. In time, they learned how to extract metals from rocks by smelting and to use the metals to make a range of products (see Figure 22.1).

The human population was only small when metal smelting was discovered and the smoke and smell from this process caused little pollution. As the human population grew, the demand for metal and other products such as pottery and glass increased. All the processing in the manufacture of these products had to be done by hand. Although there would be some pollution around the places where people gathered to make these products, the world environment was not threatened.

Figure 22.1 A Bronze Age village scene.

About 200 years ago it was discovered how machines could be used in manufacturing processes and the Industrial Revolution began. Machines could be used to produce more products than would be produced by people working on their own.

This meant that large amounts of fuel were needed to work the machines and air pollution increased (see Figure 22.2). Larger amounts of raw materials were needed and more habitats were destroyed in order to obtain them. More waste products were produced, increasing water and land pollution as the industrial manufacturing processes developed. The world population also increased, causing an increased demand for more materials which in turn led to more pollution and habitat destruction.

Figure 22.2   The smoky skyline of Glasgow in the mid 19th century.

1 What were the first materials people used?

2 Why did the pollution caused by manufacturing materials not cause a serious threat to the environment until the Industrial Revolution?

3 Why did people believe it was safe to release wastes into the environment?

For discussion

Some people believe that we must go back to the lifestyles of our earlier ancestors if the planet is to be saved. How realistic is this idea? Explain your answer.

At first, and for many years, it was believed that the air could carry away the fumes and make them harmless, and that chemicals could be flushed into rivers and the sea where they would be diluted and become harmless. Also, the ways various chemical wastes could affect people were unknown.

An awareness of the dangers of pollution increased in the latter half of the 20th century and in many countries today, steps are being taken to control it and develop more efficient ways of manufacturing materials.

The Earth's changing atmosphere

Studies from astronomy and geology have shown that the Solar System formed from a huge cloud of gas and dust in space. The Earth is one of the planets formed from this cloud. The surface of the Earth was punctured with erupting volcanoes for a billion years after it formed. The gases escaping from inside the Earth through the volcanoes formed an atmosphere composed of water vapour, carbon dioxide and nitrogen.

Figure 22.3   A smoking volcano in Indonesia.

4 How has the composition of the atmosphere changed since the Earth first formed?

5 What has changed the composition of the atmosphere?

6 The atmospheres of Venus and Mars are like the atmosphere of the Earth in the first million years of its history. What can you infer from this information?

For discussion

How is the change in the ozone layer affecting people today?

For discussion

How would our lives change if power stations could no longer supply us with electricity?

Three billion years ago the first plants developed. They produced oxygen as a waste product of photosynthesis. As the plants began to flourish in both sea and fresh water and on the land, the amount of oxygen in the atmosphere increased. It reacted with ammonia to produce nitrogen.

Bacteria developed which survived by using energy from the breakdown of nitrates in the soil. In this process more nitrogen was produced. In time, nitrogen and oxygen became the two major gases of the atmosphere. Between 15 and 30 kilometres above the Earth, the ultraviolet rays of the Sun reacted with oxygen to produce ozone. An ozone molecule is formed from three oxygen atoms. It prevents ultraviolet radiation, which is harmful to life, reaching the Earth's surface. If the ozone layer had not developed, life might not have evolved to cover such large areas of the planet's surface as it does today.

Owing to the activities of humans, the atmosphere today contains increasing amounts of carbon dioxide, large amounts of sulphur dioxide and chlorofluorocarbons (CFCs) which have destroyed large portions of the ozone layer.

Air pollution

We burn large amounts of fuel, such as coal and oil, every day in power stations to produce electricity. This provides us with light, warmth and power. The power is used in all kinds of industries for the manufacture of a wide range of things, from clothes to cars. In the home, electricity runs washing machines, fridges and microwave ovens. It provides power for televisions, radios and computers. When coal and oil are burned, however, they produce carbon dioxide, carbon monoxide, sulphur dioxide, oxides of nitrogen and soot particles that make smoke.

Carbon dioxide

Carbon dioxide is described as a greenhouse gas because the carbon dioxide in the atmosphere acts like the glass in a greenhouse. It allows heat energy from the Sun to pass through it to the Earth, but prevents

Figure 22.4 Electricity makes our lives more comfortable but its generation from fossil fuels produces pollutants.

much of the heat energy radiating from the Earth's surface from passing out into space. The heat energy remains in the atmosphere and warms it up. The warmth of the Earth has allowed millions of different life forms to develop and it keeps the planet habitable.

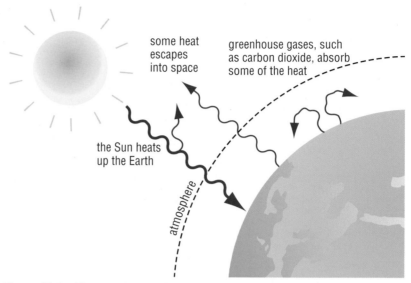

some heat escapes into space

greenhouse gases, such as carbon dioxide, absorb some of the heat

the Sun heats up the Earth

atmosphere

Figure 22.5 The greenhouse effect is important for life on Earth.

In the past the level of carbon dioxide in the atmosphere has remained fairly constant but now the level is beginning to rise. The extra carbon dioxide will probably trap more heat energy in the atmosphere. A rise in the temperature of the atmosphere will cause an expansion of the water in the oceans. It will also cause the melting of the ice cap on the continent of Antarctica and this water will then flow into the expanding ocean waters. Both of these events will lead to a raising of the sea level and a change in the climate for almost all parts of the Earth. The rise in temperature is known as global warming.

Carbon monoxide

Carbon monoxide is a very poisonous gas. It readily combines with the red pigment haemoglobin in the blood. Haemoglobin carries oxygen round the body but if carbon monoxide is inhaled, it combines with the haemoglobin and stops the oxygen being transported.

Acid rain

Sulphur dioxide is produced by the combustion of sulphur in a fuel when the fuel is burned. Sulphur dioxide reacts with water vapour and oxygen in the air to form sulphuric acid. This may fall to the ground as acid rain or snow.

Oxides of nitrogen are converted to nitric acid in the atmosphere and this falls to the ground as acid rain or as snow.

When acid rain reaches the ground it drains into the soil, dissolves some of the minerals there and carries them away. This process is called leaching. Some of the minerals are needed for the healthy growth of plants. Without the minerals the plants become stunted and may die (see Figure 22.6, overleaf).

Figure 22.6 Spruce trees in Bulgaria damaged by acid rain.

The acid rain drains into rivers and lakes and lowers the pH of the water. Many forms of water life are sensitive to the pH of the water and cannot survive if it is too acidic. If the pH changes, they die and the animals that feed on them, such as fish, may also die.

Acid rain leaches aluminium ions out of the soil. If they reach a high concentration in the water the gills of fish are affected. It causes the fish to suffocate.

Acid rain also causes chemical weathering of rocks and masonry (see page 209).

Soot and smog

The soot particles in the air from smoke settle on buildings and plant life. They make buildings dirty and form black coatings on their outer surfaces. When soot covers leaves, it cuts down the amount of light reaching the leaf cells and slows down photosynthesis.

As well as being used in industry, coal used to be the main fuel for heating homes in the United Kingdom until the 1950s. In foggy weather the smoke from the coal combined with water droplets in the fog to form smog. The water droplets absorbed the soot particles and chemicals in the smoke and made a very dense cloud at ground level, through which it was difficult to see.

When people inhaled air containing smog the linings of their respiratory systems became damaged. People with respiratory diseases were particularly vulnerable to smog and in the winter of 1952, 5000 people died in London. This tragedy led to the passing of laws to help reduce air pollution.

In Los Angeles, weather conditions in May to October lead to the exhaust gases from vehicles and smoke from industrial plants collecting above the city in a brown haze. Sunlight shining through this smog causes photochemical reactions to occur in it. This produces a range of chemicals including peroxyacetyl nitrate (PAN) and ozone. Both these chemicals are harmful to plants, and ozone can produce asthma attacks in the people in the city below.

Figure 22.7 The London smog of 1952.

7 What property of soot particles affects photosynthesis?

8 Why is carbon monoxide a deadly gas?

9 A lake is situated near a factory that burns coal. How may the lake be affected in years to come if
   a) there is no smoke control at the factory,
   b) there is no smoke control worldwide?
Explain your answers.

10 In the Arctic regions, snow lies on the ground all winter. As spring approaches and the air warms up, some of the water in the snow evaporates. Later, all the snow melts.

a) How does the evaporation of the water in the snow affect the concentrations of acids in the snow?

b) Table 22.1 shows how the pH of a river in the Arctic may change during the spring.

Table 22.1

Week	pH
1	7.1
2	7.0
3	6.9
4	6.8
5	5.5
6	5.0
7	4.7
8	5.1
9	5.5
10	5.9

i) Plot a graph of the data.
ii) Why do you think the pH changed in weeks 5–7?
iii) Why do you think the pH changed in weeks 8–10?
iv) How do you expect the pH to change in the next few weeks after week 10? Explain your answer.

11 Does a tall chimney solve the problem of air pollution? Explain your answer.

The danger of lead

A combustion reaction takes place inside car engines. In this reaction, petrol is burned to release energy to push the pistons in the engine. In the past, lead was added to all petrol to improve the combustion reaction and the engine's performance. The exhaust gases carried the lead away as tiny particles. They were inhaled by people and settled on their food and skin. Lead in high concentrations in the body causes damage to the nervous system including the brain. Children absorb lead into their bodies more readily than adults and in areas of cities where there are large amounts of exhaust gases from cars, high levels of lead have been found in children's blood.

Improving air quality

The air around a factory can be kept clean by using a tall chimney to release the smoke high into the air (see Figure 22.8). Winds take the smoke away from the factory and its surrounding area. The harmful constituents in factory smoke can be removed chemically and physically.

Figure 22.8   A factory with a very tall chimney, in the north of England.

Chemical removal of sulphur dioxide

Sulphur dioxide can be removed in two ways to form useful products.

Lime can be sprayed into the waste gases where it combines with sulphur dioxide to form calcium sulphate. This rocky material can be used in making the foundation layer of roads.

Ammonia can be mixed with waste gases where it reacts with sulphur dioxide to form ammonium sulphate, which can be used as a fertiliser.

12 Treating waste gases is expensive. Calcium sulphate can be used as a building material in road making and ammonium sulphate can be used as a fertiliser.

  a) How might treating waste gases affect the price of the product being made?

  b) How might a company earn extra money after it has fitted equipment to treat waste gases? How will this affect the price of the product?

13 The middle of a catalytic converter has a honeycomb structure. Why is this structure used?

For discussion

What can we do to reduce air pollution?

Physical removal of particles

Most of the particles in smoke have a small charge of static electricity. The particles can be removed by a device called an electrostatic precipitator. This device has highly charged metal plates. When the smoke passes through the precipitator, the particles are attracted to the plates and the remaining gases pass on.

Smokeless fuel

Substances which cause harmful smoke can be removed from fuel before it is used. Coal, for example, can be heated without air to remove the tars and gas which make the coal burn with a smoky flame. Coal treated in this way forms the fuel called coke.

Unleaded petrol

Engines have been developed which run on unleaded petrol, yet still give good performance.

Catalytic converters

Many cars are now fitted with a catalytic converter. This device forms part of the exhaust system. Inside the converter is a catalyst made of platinum and rhodium. The waste gases from the engine take part in chemical reactions in the converter which produce water, nitrogen and carbon dioxide.

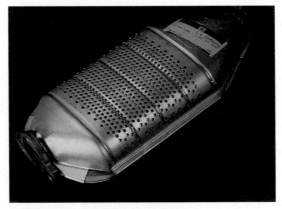

Figure 22.9 A catalytic converter.

CFC replacements

CFCs have been replaced by carbon dioxide and hydrocarbon gases in the manufacture of aerosols and refrigerators in many countries.

Weathering

When a surface of a rock is exposed to the air, the process of weathering begins. In this process the surface of the rock is broken up and the fragments that are made fall away. This exposes a new surface of the rock and the process of weathering begins again.

The action of weathering can be studied in a graveyard. Often a large number of gravestones are made from the same stone, such as one which occurs locally. If the gravestones are of many different ages they can be used for studying the effect of weathering on one type of stone

14 How could you compare the effect of weathering on gravestones made from different rocks?

over a long period of time. It should be found that on the newest gravestones the surface is smooth and the lettering very clear. On the oldest gravestones the surface may be rough and crumbly and the letters worn away and difficult to read (see Figure 22.10).

There are two kinds of weathering processes – chemical weathering and physical (or mechanical) weathering. We are concerned here with chemical weathering.

Figure 22.10   The effect of weathering on gravestones.

Chemical weathering

Chemical weathering occurs when a rock surface is broken down by a chemical reaction.

The effect of water

If you have extracted salt from rock salt you will have seen how water can dissolve the salt in the rock. Salt is made from a mineral called halite. It is unusual because it will dissolve in water. Most other minerals do not dissolve in water, but if the water is acidic they may take part in a chemical reaction which produces weathering.

The effect of acid rain

When water vapour condenses on dust particles to form raindrops it is pure water but a change soon takes place. Carbon dioxide, which is present in the air, dissolves in the water and forms carbonic acid. The word equation for this reaction is:

carbon dioxide(g) + water(l) $\rightleftharpoons$ carbonic acid(aq)

The double arrow shows that the reaction is a reversible reaction. It is divided into two parts – the forward reaction reads from left to right, and the back reaction reads from right to left. The speed at which carbon dioxide and water form the acid is greater than the speed at which the acid breaks up to form its reactants. This means that the water in the raindrop becomes acid.

The formation of acid rain from carbon dioxide occurs naturally but acid rain can be produced by pollution (see page 205).

Figure 22.11   Halite.

Acid rain and limestone

Limestone is made from calcium carbonate. When water containing carbonic acid runs into cracks in the rock this reaction takes place:

$$\text{carbonic acid(aq)} + \text{calcium carbonate(s)} \rightarrow \text{calcium hydrogencarbonate(aq)}$$

The calcium hydrogencarbonate dissolves in the water and is washed away from the reaction site.

The surface of a large piece of limestone rock, in time, breaks up to form small cracks called grikes with pieces of rock between them called clints. The surface of clints and grikes is known as a limestone pavement. Deep below a limestone pavement there may be caves, produced by streams flowing through the rock. The acidic water of the stream has broken down large quantities of rock to make huge hollows. In some places the roof of a huge cave has collapsed to make a gorge.

Figure 22.12 A limestone gorge (left) and a limestone pavement (right).

The effect of climate

The climate of a place is a description of the weather that usually occurs there every year. Two features of the climate that are important for chemical weathering are the rainfall and the temperature.

Water is needed for chemical weathering to take place and heat speeds up chemical reactions. Therefore chemical weathering of rocks is much greater in a place with a hot, wet climate, such as in rainforest regions, than a place with a cold, dry climate, such as the interior of the Antarctic continent.

> **15** How great will chemical weathering be in a desert? Explain your answer.

Water pollution

Fresh water

Fresh water, such as streams and rivers, has been used from the earliest times to flush away wastes. Over the last few centuries many rivers of the world have been polluted by a wide range of industries including textile and paper making plants, tanneries and metal works. People in many countries have become aware of the dangers of pollution and laws have been passed to reduce it. Ways have been found to prevent pollution occurring and to recycle some of the materials in the wastes.

The most harmful pollutants in water are the PCBs (polychlorinated biphenyls) and heavy metals such as cadmium, chromium, nickel and lead. In large concentrations these metals damage many of the organs of the body and can cause cancers to develop. PCBs are used in making plastics and, along with mercury compounds, are taken in by living organisms at the beginning of food chains (see page 32). They are passed up the food chain as each organism is eaten by the next one along the chain. This leads to organisms at the end of the food chain having large amounts of toxic chemicals in their bodies which can cause permanent damage or death.

Figure 22.13   This cellulose factory is causing the water to become polluted.

The careless use of fertilisers allows them to drain from the land into the rivers and lakes and leads to the overgrowth of water plants. When these die, large numbers of bacteria decompose the plants and as they do so, the bacteria take in oxygen from the water. The reduction in oxygen levels in the water kills many water animals. Phosphates in detergents also cause an overgrowth in water plants which can lead to the death of water animals in the same way.

Figure 22.14   The excessive use of fertilisers leads to algal bloom in rivers and kills fish.

16 How are the lives of people who live by polluted rivers and catch fish from them put at risk?

17 The water flowing through a village had such low levels of mercury in it that it was considered safe to drink. Many of the villagers showed signs of mercury poisoning. How could this be?

Figure 22.15 Oil spills like this can have a huge effect on the sea and coastal wildlife.

Sea water

The pollutants of fresh water are washed into the sea where they may collect in the coastal marine life. The pollutants may cause damage to the plants and animals that live in the sea and make them unfit to be collected for human food.

Large amounts of oil are transported by tankers across the ocean every day. In the past the tanker crew flushed out the empty oil containers with sea water to clean them. The oil that was released from the ship formed a film on the water surface which prevented oxygen entering the water from the air. It also reduced the amount of light that could pass through the upper waters of the sea to reach the phytoplankton and allow them to photosynthesise.

The problem of this form of oil pollution has been reduced by adopting a 'load on top' process, where the water used to clean out the containers is allowed to settle and the oil that has been collected floats to the top. This oil is kept in the tanker and is added to the next consignment of oil that is transported.

Occasionally a tanker is wrecked. When this happens large amounts of oil may spill out onto the water and be washed up onto the shore (see Figure 22.15). This causes catastrophic damage to the habitat and even with the use of detergents and the physical removal of the oil the habitat may take years to recover.

Pollution and destruction of the land environment

The major chemical pollutants on land are pesticides which can affect human health, and radioactive chemicals accidentally released from nuclear power plants which can cause cancer to develop. DDT is a pesticide that causes serious long-term problems (see pages 117–118) and it is now banned in many countries. However, it is still used in some countries where no laws exist to restrict its use. They continue to use it because it is very effective in controlling the mosquitoes that spread malaria.

> **18** How does oil floating on the surface of the sea affect the organisms living under it?

Figure 22.16   A tip with a methane 'breather'.

The discarded products of manufacturing industries produce a pollution problem in every country. The tips in which the waste is stored take up space.

Today, many tips are carefully filled so that when they are full they can be covered with soil and new habitats established on top of them. While the rubbish is settling and decomposing on the tip some of it gives off methane gas. This can be collected by a system of pipes and used as a fuel.

Most raw materials have to be taken out of the ground. In some cases mine shafts are sunk into the ground and the material is removed with little damage to the surrounding habitat. Lead, zinc and some copper and coal are mined in this way.

In open cast mining, the land surface is removed to extract the raw material (see Figure 22.17). Aluminium and some coal and copper are removed like this. It causes complete habitat destruction. If this occurs in rainforest areas the forest may not be able to grow back again when the mining operation is over because the thin layer of soil on which the forest grew may have been completely washed away.

19 How do the methods of extracting raw materials affect plants and animals that live in the same area?

Figure 22.17   Open cast mining in a rainforest in South America.

Renewable and non-renewable materials

Raw materials can be divided into two groups – renewable materials and non-renewable materials. Wood is an example of a renewable material. As trees are cut down to provide the raw material for wood products, young trees are planted to replace them. Iron is an example of a non-renewable material. There is a certain amount of it in the Earth's crust which is not replaced after it is used up. As the iron ore is mined the supply left in the ground is reduced. In time there could be none left to use.

As the human population increases, the demand for raw materials also increases. Although renewable materials can be replaced, the extra demand means that extra space has to be found for the material to be re-formed. This can result in habitat destruction. An example of this is where moorlands are planted with forests of fast growing trees to be used in manufacturing.

As non-renewable raw materials cannot be replaced, studies have been made to find out how much of each material is left on the Earth. For the purpose of the study the raw materials are divided into three groups. These are the stocks, the reserves and the resources. The stocks are the materials which are already mined and stored ready for use. The reserves are materials still in the ground that can be mined economically (they are not too expensive to mine). The resources cannot be mined economically – they are too expensive to mine at present. (Note that another use of the word resource in science is to mean a supply that is readily available.) Once stocks are used up, reserves will be mined and converted into stocks. Resources may then become reserves and a material comes closer to being used up. This process can be slowed down so the material is conserved by recycling.

Figure 22.18 A coniferous forest being replanted.

Recycling

The products into which materials are made are often used only for a certain length of time. A newspaper may be read for a day, a bottle of lemonade may last three days, an item of clothing may last a year and a car may last 15 years. If the products are thrown away when their use is over, the materials in them just stay in the ground in a tip. They take up space and have to be replaced by extracting more raw materials and using large amounts of energy in the manufacturing processes. Recycling the materials saves space, raw materials and energy.

Paper is made from smashing wood into a pulp of tiny fibres and then binding them together in a thin sheet. When paper is recycled it is made into a pulp of fibres again, without having to use energy and chemicals to break down the wood.

---

20 Many natural forests have a mixture of many different species of tree. They are of different ages and are irregularly spaced out. Many planted forests have very few tree species. The trees are the same age and are regularly spaced out.

  a) In what ways are the planted forests different from the natural forests?

  b) Do you think the two forests will support the same wildlife? Explain your answer.

21 Assume the world stock of a material is 10 000 000 tonnes and it is used at a rate of 250 000 tonnes a year.

  a) How long will the stocks last?

  b) When will the stocks run out?

22 Imagine that a new product has been invented that uses the material in question 21. An extra 30 000 tonnes a year of the material is extracted for this product.
  a) How long will the world stocks now last?
  b) When will the stocks now run out?

23 Imagine a recycling programme has been set up in which 200 000 tonnes of the material in question 21 could be recycled each year.
  a) How long would the stocks now last using:
    i) 250 000 tonnes a year,
    ii) 280 000 tonnes a year?
  b) What effect does the recycling programme have on the reserves of the material?

24 If 1000 million tonnes of bauxite are mined every year and it is estimated that stocks will last until about 2240, how much bauxite is on the Earth?

25 What are the benefits of recycling?

26 When oil and natural gas supplies are used up, do you think the stocks of coal will still be expected to last until 2300? Explain your answer.

27 How does the recycling of materials affect the stocks of fossil fuels? Explain your answer.

Glass is made from sand, limestone and soda, and a large amount of heat energy is required. Less energy is needed to melt recycled glass and make it ready for use again. The recycled glass is mixed with the raw ingredients as new glass products are made.

Large amounts of energy are needed for the extraction of metals such as iron and aluminium. Less energy is needed to melt them down than to extract new metals from their ores. By recycling metals, less fuel is used and the stocks of the ores are conserved.

Methods of separation

Materials for recycling can be separated by people and taken to recycling centres (see Figure 22.19) or they can be separated after the collection of refuse. The magnetic separator (see Figure 17.7, page 156) is used to separate iron and steel from other materials.

In industry, products which are wastes in one process can be collected and used elsewhere. For example, in the purification of copper the metals silver and gold are produced. These metals are not discarded but sold to people such as jewellery manufacturers who can use them.

Some of the reactions which take place in the chemical industry produce heat energy. This is not released but used in other parts of the chemical plant. For example, the heat produced when sulphur and oxygen combine in a combustion reaction is used to melt the solid sulphur at the beginning of the process to manufacture sulphuric acid.

Figure 22.19 Recycling centre.

Materials and energy

The processing of all materials needs energy and this is provided mainly by the fossil fuels – coal, oil and natural gas. These are non-renewable raw materials and while stocks of coal may only last until the year 2300, stocks of oil and natural gas are predicted to be used up in your lifetime, if used at the present rate. When materials are recycled there is a reduction in the amount of energy used to make the new products. Although some energy is used in the recycling process, it is usually less than the energy used in extraction.

Using materials in the future

Increasing amounts of many materials are being recycled and new ways are being found to save energy in chemical processing to meet the demands of the human population, today and in the future.

New materials are made every year through investigations into the way different chemicals react together. From these discoveries, materials are selected which can perform a task more efficiently than an existing material and require smaller amounts of raw materials and energy. In the long term, there are plans to set up mines on the Moon to extract minerals and to process them, to make materials in space for use on Earth and in further space exploration.

Figure 22.20   Impression of a manned lunar base.

Summary

- From the earliest civilisations human activity has caused some pollution but the problem greatly increased with the Industrial Revolution (*see page 202*).
- Air is polluted by solid particles and by chemicals (*see page 204*).
- There are a number of ways in which air quality can be improved (*see page 207*).
- Chemical weathering occurs when a rock surface is broken down by chemical reactions, for example by the action of acid rain (*see page 209*).
- Fresh water may be polluted with dangerous heavy metals (*see page 210*).
- Sea water may be polluted with oil (*see page 212*).
- Some pesticides, rubbish and the extraction of raw materials damage the land environment (*see pages 212–213*).
- Some materials are renewable while others are non-renewable (*see page 214*).
- Recycling conserves stocks of raw materials, including fuels (*see page 214*).

End of chapter question

1 Using the information in this chapter, what policies would you suggest to the governments of all countries to improve the quality of the world environment and to ensure that future generations have the resources they need to meet their needs?

physics

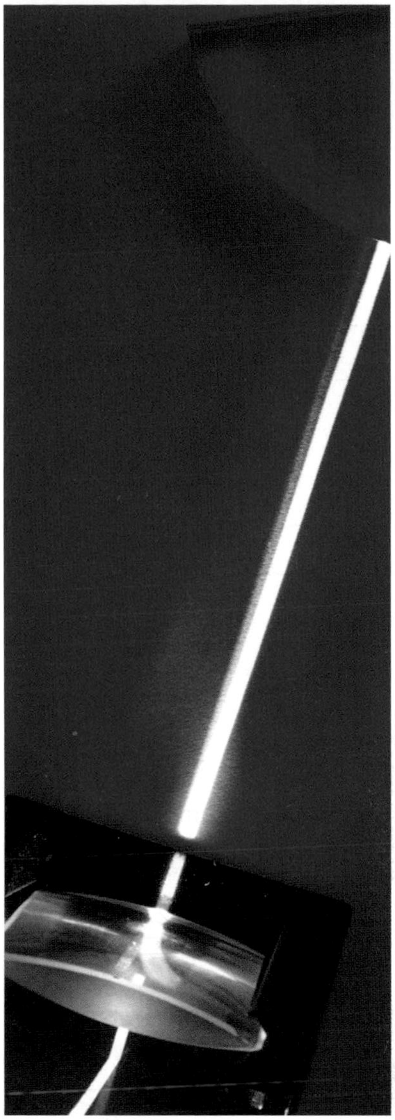

Figure 24.3   When the objects fall their stored gravitational potential energy is released.

Strain potential energy

Some materials can be easily squashed, stretched or bent, but spring back into shape once the force acting on them is removed. They are called elastic materials. When their shape is changed by squashing, stretching or bending they store energy, called strain energy or spring energy, which will allow them to return to their original shape. This form of stored energy is also called elastic potential energy.

   A spring stores energy when it is stretched or squashed. Gases store strain energy in them when they are squashed. For example, when the gas used in an aerosol is squashed into a can it stores strain energy. Some of this is used up when the nozzle is pressed down and some of the gas is released in the spray.

> **6** Look at Figure 24.4. When is elastic potential energy stored and when is it released in
> **a)** a toy glider launcher,
> **b)** the elastic cords or springs beneath a sun lounger,
> **c)** a diving board?

Figure 24.4   Places where strain energy can be stored and released.

Nuclear energy

Every substance is made from atoms and at the centre of each atom is a nucleus. It is made from particles called protons and neutrons. In the atoms of most elements these particles are bound strongly together. These atoms are said to be stable atoms. In some large atoms the nuclear particles are not bound strongly together and the unstable nuclei may

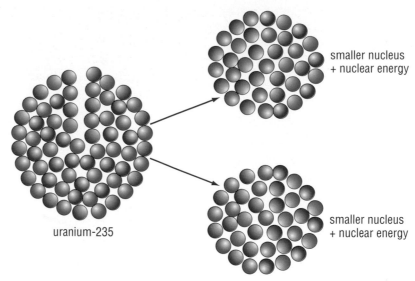

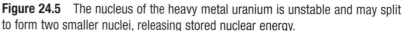

smaller nucleus + nuclear energy

smaller nucleus + nuclear energy

uranium-235

Figure 24.5 The nucleus of the heavy metal uranium is unstable and may split to form two smaller nuclei, releasing stored nuclear energy.

7 What is the difference between nuclear fission and nuclear fusion?

split up to form different, smaller nuclei which are more stable. As the unstable atoms form more stable atoms they release energy. This process of releasing stored nuclear energy is called nuclear fission.

In the Sun and other stars nuclear energy is released instead by nuclear fusion. The gravitational forces are very strong at the centre of the star and the hydrogen nuclei collide at such high speeds that they actually join together (fuse) to form helium nuclei. During this process large amounts of energy are released. Most of this energy escapes from the star as heat and light.

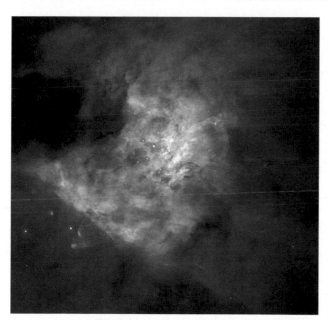

Figure 24.6 These young stars shine because of nuclear fusion.

Other forms of stored energy

Two other forms of stored energy are magnetic potential energy – the energy of a magnetic field, and electrostatic potential energy – the energy of an electric field.

Work and stored energy

Energy may be stored as a result of work being done. For example, the child at the top of the slide in Figure 24.3 has gravitational potential energy because of the work she has done in climbing to the top. She had to do work against the force of gravity.

Similarly, a stretched elastic band has stored strain energy because of the work done on it in stretching it against the strain forces.

For discussion

A nail held close to a magnet has magnetic potential energy. Why is this so?

What happens to the magnetic potential energy of the nail when it is released close to the magnet?

Kinetic energy

Kinetic energy is also known as movement energy. When an object with kinetic energy touches another object, it exerts a contact force which may set the second object moving too.

Any moving object has kinetic energy. The object may be as large as a planet or as small as an atom or an electron.

8 Look out of a window and make a list of everything you can see that has kinetic energy.

Figure 24.7 The kinetic energy of the demolition ball is transferred to the building and breaks it up.

Internal energy

In fact, you could have listed everything you can see when answering question 8, since all the atoms and molecules which form solids, liquids or gases have kinetic energy and move at random. The atoms in a solid move the least. Each has a position in the solid structure about which it vibrates to and fro. The atoms in a liquid move around and past each other and the atoms in a gas can move freely in all directions. This movement of the atoms inside a solid, liquid or gas is due to the internal energy of the substance. The internal energy is increased by heating a substance. This increases the movement and separation of the atoms.

9 How can the internal energy of an object be reduced?
10 How does reducing the internal energy affect the atoms inside the object?

Thermal energy

The energy supplied to a substance which increases its internal kinetic energy, or the energy lost from a substance which decreases its internal kinetic energy, is commonly called heat, or heat energy. The scientific term is thermal energy. The process by which thermal energy moves is called thermal energy transfer or thermal transfer (see page 283).

Sound energy

Sound energy is produced by the vibration of an object such as the twang of a guitar string. The energy passes though the air by the movement of the atoms and molecules. They move backwards and forwards in an orderly way. This makes a wave that spreads out in all directions from the point of the vibration. Sound energy can also pass through solids, liquids and other gases. The atoms move in a similar way to the turns on a slinky spring when a 'push-pull' wave moves along it.

to-and-fro vibration of the turns as the push-pull wave passes

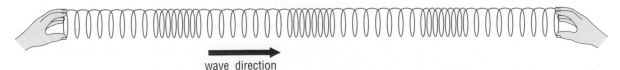

wave direction

Figure 24.8 A slinky shows how sound waves move.

Electrical energy

Electrical energy is due to the movement of electrons though an electrical conductor such as copper or graphite when it is part of a circuit with a power supply (such as a battery) and the switch is closed.

Radiation energy

This kind of energy travels in waves that have some properties of electricity and some properties of magnetism. They are called electromagnetic waves. There is a huge range of possible wave sizes, or wavelengths. Electromagnetic waves are split into seven groups according to wavelength, as Figure 24.10 shows. The different groups have different properties and different uses. The two most familiar groups are light and radio waves.

11 In what ways is electrical energy put to work in your home?

12 Which radiation energy has
  a) the longest wavelength,
  b) the shortest wavelength?
13 Which radiation energy can our eyes detect?

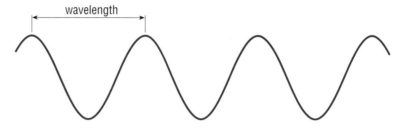

Figure 24.9   A wave showing wavelength.

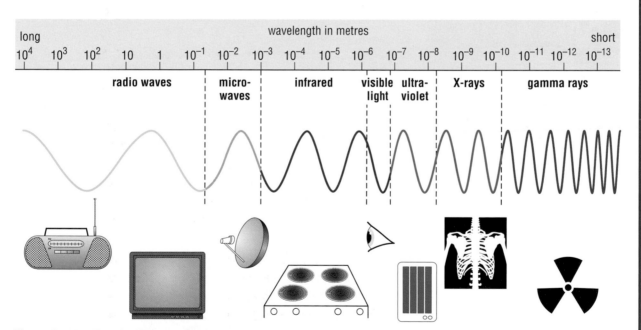

Figure 24.10   The electromagnetic spectrum.

Light energy

The light energy escaping from the Sun can be spread out by a prism or a shower of raindrops into light of different wavelengths. This forms the colours of the rainbow (see page 304) because our eyes see different wavelengths of light as different colours.

Infrared radiation

Infrared radiation carries heat (thermal) energy in the form of electromagnetic waves which we detect as warmth on our skin.

Ultraviolet light

Ultraviolet or UV light causes chemical changes in our skin that give the skin a tan. The excessive amounts of ultraviolet light reaching the Earth because of the destruction of parts of the ozone layer (see page 204) may also stimulate the growth of skin cancers.

Energy changes

Energy can be transferred from one form to another. When an amount of energy is transferred none is lost. There is the same amount of energy afterwards as there was before. This constancy of energy is known as the Law of the Conservation of Energy.

The object or material in which the energy changes form is called the energy converter or energy transducer. The transfer of energy can be written using this format:

energy input → *energy transducer* → energy output

For example, when a candle burns the energy transfer is:

chemical energy → *candle* → light energy
→ heat energy

When you run the energy transfer is:

chemical energy in food → *human body* → kinetic energy
→ heat energy

The flow of energy through one or more energy transducers is called an energy chain. For example, the energy chain produced when a motor on an electric golf buggy is switched on and moves the vehicle is:

chemical → *electrical* → electrical → *motor* → kinetic + sound + heat
energy   *cell*   energy   energy

For discussion

Is radiation energy harmful? Explain your answer.

Figure 24.11 A candle is an energy transducer.

Figure 24.12 People are energy transducers.

14 What are the energy transfers when
   a) a dog barks,
   b) a television set is switched on?
15 What energy transfer takes place
   a) in the Sun,
   b) when plants make food,
   c) when you run after a meal?
16 Link your answers to questions 15a), b) and c) together to make an energy chain.
17 Make three more energy chains which feature at least three energy transducers.

Figure 24.13 There are several energy transfers going on in a device like an electric golf buggy.

A closer look at energy changes

The energy chains just considered have shown the main transfers in energy but if we considered them in more detail we would find that other transfers also take place. Here are the ones that take place when a ripe apple falls off a tree.

An apple on a tree has gravitational potential energy. When its stalk weakens and breaks, the apple's weight pulls it down and its stored energy becomes kinetic energy. As the apple falls through the air it collides with atoms in the air and causes them to move faster – the air is warmed a little. This means that some of the apple's energy is transferred to heat energy. When the apple strikes the ground a thud is heard. This means that some of the energy is transferred to sound energy. The apple and the ground are warmed slightly by the collision, so more of the apple's kinetic energy is transferred to heat energy, which increases the internal energy of the apple and the ground.

The energy transfers that occurred when the apple fell to the ground can be written as:

gravitational potential energy → kinetic energy

→ heat energy to air

→ heat energy to ground and apple

→ sound energy

In fact, whenever energy changes, *some of it is always lost as heat (thermal) energy.*

18 Look back at the energy chains you made in answer to questions 16 and 17. See if you can add more detail to the energy changes.

For discussion

What energy changes occur when you jump up and down?

Sources of energy

We use many different sources of energy to provide everything we use in our everyday lives. The sources of energy may be as close as the wind in your face or as distant as the inside of the Sun, 150 million kilometres away.

On the following pages the sources of energy have been divided into three groups as shown in Table 24.1.

Table 24.1 Sources of energy.

Nuclear fusion in the Sun	Nuclear reactions on Earth	Gravitational forces between Earth, Moon and Sun
produces **a)** light energy which is changed into: • biomass • fossil fuels • electrical energy by solar cells	produce internal (heat) energy from **a)** controlled fission reactions of minerals from the Earth's crust, e.g. uranium	produce tidal energy
b) infrared (heat) energy which is changed into: • movement energy in the atmosphere used to drive the water cycle • movement energy in water waves • internal energy of water in solar panels	**b)** naturally occurring nuclear reactions in radioactive rocks below the Earth's crust; this internal energy is called geothermal energy	

Nuclear fusion in the Sun

Figure 24.14   A close-up of the Sun's surface.

The Sun is a star, a huge fireball of hydrogen and helium gases which is 1 392 000 kilometres in diameter. Over one million planets the size of the Earth could fit inside the Sun. At the centre of the star it is so hot that hydrogen nuclei move fast enough to hit each other and fuse together to produce helium. As this happens energy is released which is radiated from the Sun's surface as electromagnetic waves, mostly light and infrared radiation. It is believed that the Sun will continue to release energy in this way for another 5000 million years. The Sun is our main source of energy.

Energy from the Sun's light

Biomass

Biomass is the amount of matter in a living thing. Plant biomass is produced by plants from the raw materials of carbon dioxide, water and minerals.

Green plants possess a pigment called chlorophyll which traps some of the energy in sunlight. The plants use this energy to join chemicals together to make food which is then used to build up the plant's body or to take part in processes in the plant which keep it alive.

Food

The food stored in plants may be eaten by herbivores, such as sheep, and omnivores such as ourselves. We have bred certain plants for cultivation as crops which provide us with food rich in energy. Food provides animals with chemical potential energy. They use this energy to keep alive and build up their own bodies. We also take in stored chemical energy from animals when we eat their meat.

The unit for measuring energy in food is the kilojoule. See Table 5.3, page 47, for the energy values of 100 g of some common foods.

Fuel wood

The energy stored in plants can be released by drying them and setting fire to them. In many parts of the world, particularly in developing countries where there is no coal, wood is used as a fuel for cooking and heating.

> **19** Why does the Sun release energy?

> **20** Why does food from
> **a)** plants and
> **b)** animals contain energy?

Figure 24.15   In Nepal, ovens are fuelled by wood.

Figure 24.16   This pump delivers alcohol petrol.

Ethanol

Some countries, such as Brazil, have few natural supplies of oil from which to make petrol. A shortage of fuel for vehicles is prevented by fermenting plants such as sugar cane and grain to make ethanol, a type of alcohol. This is mixed with petrol and reduces the demand for oil.

Energy from food wastes

Biogas

The wastes produced when food has been eaten can be used as a source of energy. Vegetable waste and the wastes from humans and animals can be stored in biogas digesters, where microorganisms feed (see Figure 24.17). As they feed, methane gas is produced. This gas can be used for cooking or lighting and as a fuel to generate electricity.

In sewage works the solid wastes are stored in digesters where they form the food for microorganisms. The methane gas produced can be used as a source of heat energy to help in the sewage treatment.

Dung

In countries without coal and where fuel wood is scarce the dung of domestic animals, such as cattle, is collected, dried, made into fuel cakes for burning (see Figure 24.18).

Figure 24.17   Biogas digester in Egypt.

Figure 24.18   Making fuel cakes from animal dung.

Fossil fuels

Coal is formed from large plants which grew in swamps about 275 million years ago. These plants used energy from sunlight in the same way as plants do today. When they died they fell into the swamps. There was a lack of oxygen in the swamp water which prevented bacteria growing and decomposing the dead plants (see page 77). Eventually the plants formed peat. Later the peat became buried and was squashed by the rocks that formed above it. The increase in pressure squeezed the water out of the peat and warmed it. These processes slowly changed the peat into coal.

Tiny plants and animals live in the upper waters of the oceans and form the plankton. When they die they sink to the ocean floor. Over 200 million years ago the dead plankton which collected on the ocean floor did not decompose because there was not enough oxygen there to allow bacterial decomposers to grow. The remains formed a layer which eventually became covered by rock. The weight of the rock squeezed the layer and heated it. This slowly converted the layer of dead plankton into oil and methane gas. This is the gas that is supplied to homes as natural gas. Several fuels are obtained from oil.

21	What conditions helped fossil fuels to form?

Figure 24.19   The three fossil fuels.

Solar cells

A solar cell is made from a material which converts some of the energy in sunlight into electrical energy. Solar cells are used on satellites, space probes and space stations to provide power to work the machines on board.

Figure 24.20   Arrays of solar cells project from the International Space Station to collect some of the Sun's energy.

Energy from the Sun's heat

Most of the energy from the Sun is in the form of infrared radiation. This warms the Earth and its atmosphere.

Movement of the atmosphere

The air nearest the Earth's surface warms the fastest and then rises. It is replaced by cooler air. The cycle repeats itself and a convection current is set up (see page 284). Various circulations of this type occur in the Earth's atmosphere. Regions of high and low pressure develop in the atmosphere which in turn lead to the production of winds. The kinetic energy in wind can be used to turn the blades of a wind turbine and produce electricity.

Figure 24.21 Wind turbines on a wind farm.

Movement of water

Flowing water

The Sun's heating effect causes the evaporation of water from the surfaces of the oceans and lakes. The water vapour that is produced rises in the warm air then condenses to form clouds as the air cools. Winds blow the clouds around the globe, and further falls in temperature over land or sea make the clouds produce rain. The water falling on land forms rivers which flow back to the oceans and complete the water cycle (see page 133).

The energy in river water can be used in the following way. If a dam is built across the river, the water collecting behind it has stored gravitational energy. When the water is released from the lower part of the dam the high pressure there makes it flow very fast. It therefore has a large amount of kinetic energy and this can be used to spin a turbine and generate electricity.

Figure 24.22 Array of turbines in a hydroelectric power station in Russia.

Figure 24.23 This machine (here being towed back from sea trials) converts the energy in the up-and-down motion of passing waves into the turning motion required to make an electric generator work.

Waves

The winds blow on the surfaces of the oceans and produce waves. The kinetic energy in the wind is converted into kinetic energy of the water. Machines are being developed to use the energy in waves to generate electricity.

Solar panels

A solar panel is designed to heat water in a house in sunny weather. In the panel are pipes which carry water. They are connected to a coiled pipe in the domestic hot water tank. The pipes in the panel run over a black surface which absorbs infrared radiation from the Sun. The panel is double glazed to reduce the loss of heat (thermal) energy from around the pipes. The hot water produced in a solar panel on a sunny day is circulated through the coil in the hot water tank to warm the water there.

Figure 24.24 These roofs in Sweden are fitted with solar panels.

> **22** Make a chart showing the relationship between the Sun and the different energy sources described so far.

Energy from nuclear reactions on Earth

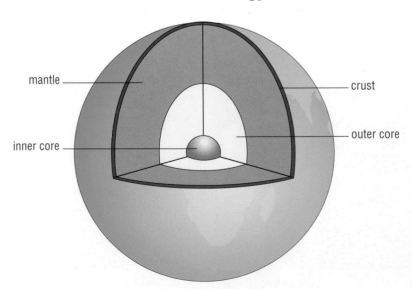

mantle

inner core

crust

outer core

Figure 24.25 The structure of the Earth.

Almost all the elements which form the matter from which the Earth is made have been made by nuclear fusion in stars in the distant past. Some elements are unstable and 'decay' in nuclear reactions to form more stable elements. The unstable elements are called radioactive elements. When they undergo radioactive decay they release particles and energy.

We use energy from radioactive materials which are found below the Earth's crust. The structure of the Earth is shown in Figure 24.25.

Figure 24.26 The energy to heat a geyser is provided by hot rocks below ground.

These radioactive materials release energy that heats the core and mantle of the Earth. The heat also passes into the crust. In some places it makes water in the rocks heat up and form geysers as shown in Figure 24.26.

The heat in the rocks of the crust is used as a source of energy in some countries, for example New Zealand, Iceland and some parts of the USA. Cold water is pumped down to the hot rocks where it is turned to steam. This returns through pipes to the surface where it is used to generate electricity in a power station.

Figure 24.27 A geothermal power station.

Energy from gravitational forces

The Moon, the Sun and the tides

The water in the oceans is not only pulled down by the Earth's gravity but is also affected by the gravitational pull of the Moon, and to a smaller extent by the gravitational pull of the Sun. Differences in the force of attraction of the Moon on different parts of the Earth cause the water beneath the Moon and on the opposite side of the Earth to rise up slightly, causing a high tide. The level of water in other parts of the ocean falls, and produces a low tide at those places.

High tide
Figure 24.28

Low tide

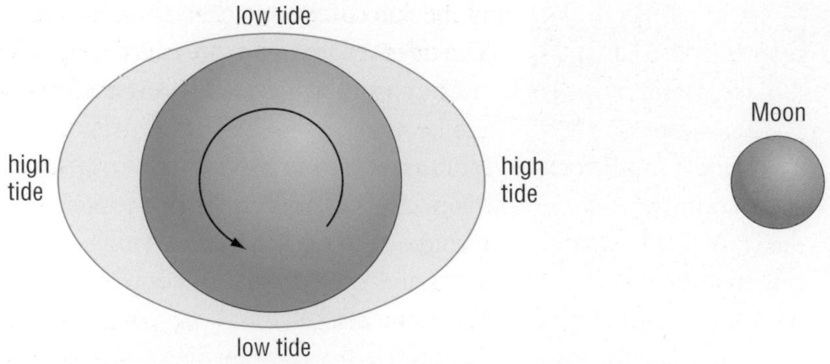

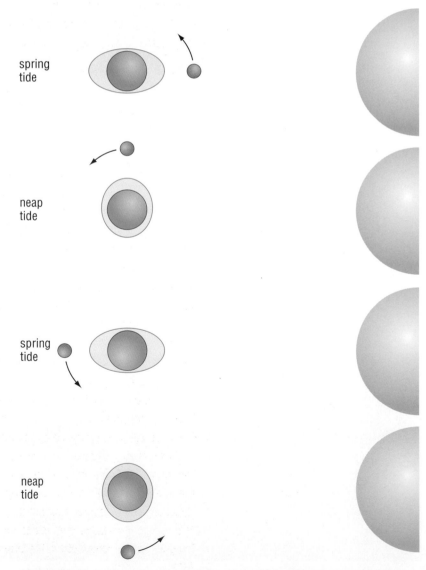

low tide

high tide

high tide

low tide

Figure 24.29 The attraction of the Moon produces high and low tides.

Moon

As the Earth rotates, different regions of the seas and oceans move under the Moon and experience the full force of its gravitational pull. This brings a period of high tide to that region which is followed by a period of low tide as the Earth continues its rotation.

Twice a month the Sun and the Moon are in line as the Moon orbits the Earth (see Figure 24.30). At these times the difference in the sea level between high and low tides is greatest as the small gravitational pull of the Sun reinforces the effect of the stronger pull of the Moon. The tides at these times are called spring tides.

When the Moon is furthest out of line with the Sun and the Earth, the difference between high and low tides is least since the gravitational pull

23 The Sun is much larger than the Moon yet its gravitational pull on the oceans is less than that of the Moon. Why is this?

spring tide

neap tide

spring tide

neap tide

Figure 24.30 How the positions of the Sun and the Moon cause a change in the tides.

Figure 24.31 The La Grande I project in Canada generates electricity from the tides.

of the Sun counteracts that of the Moon. The tides at these times are called neap tides.

The energy stored in the water at high tide can be used to generate electricity. A dam is built across an estuary and the rising water of the tide passes through pipes in which turbines can turn. This movement is passed to a generator and electricity is produced. When the tide falls the water passes through the pipes in the opposite direction and turns the turbines again, so more electricity can be generated.

Non-renewable and renewable energy sources

The sources of energy can be divided into two groups according to whether they will eventually be used up or whether they can be constantly replaced.

Non-renewable energy sources
These sources cannot be replaced once they have been used up. These energy sources are the fossil fuels and radioactive materials.

Renewable energy sources
These sources can be replaced. They are light and heat from the Sun, biomass, geothermal energy and kinetic energy of the wind, water, waves and the tides.

Summary

- Energy allows something to do work (*see page 228*).
- Energy and work are measured in joules (*see page 228*).
- Energy can be stored. Examples of stored energy are chemical energy, gravitational potential energy, elastic potential (strain) energy and nuclear energy (*see pages 229–231*).
- Movement energy is known as kinetic energy (*see page 232*).
- Internal energy in a substance is due to the movement of its atoms and molecules (*see page 232*).
- Thermal energy is the correct scientific term for heat energy. When thermal energy is transferred, the internal kinetic energy of a substance increases (*see page 232*).
- Sound energy is transferred by waves in which atoms move backwards and forwards (*see page 232*).
- Electrical energy is produced by the movement of electrons (*see page 233*).
- Radiation energy is transferred by electromagnetic waves (*see page 233*).
- Energy can change from one form into another (*see page 234*).
- The Sun is the major source of energy for our planet (*see page 235*).
- Radioactive materials release energy (*see page 240*).
- Heat energy is released from inside the Earth (*see page 240*).
- The gravitational forces of the Moon and the Sun cause the tides (*see page 241*).
- Energy sources can be renewable or non-renewable (*see above*).

End of chapter questions

A group of pupils was investigating the gravitational potential energy in a 15 cm long nail. They suspended it above a block of clay, measured the distance to its tip then let it go. The pupils measured the depth to which the nail sank in the clay. Table 24.2 shows their results for four experiments.

Table 24.2

Height of nail above clay (cm)	Depth of indent (cm)
25	0.9
50	1.6
75	2.3
100	3.0

1 How do you think they measured the depth of the indent in the clay?

2 Plot a graph of their results.

3 How could you use the graph to predict the indent made by the nail from a height greater than 1 metre? Give an example.

A second group of pupils investigated the gravitational potential energy of a brass sphere which was dropped from different heights into soft clay. They measured the diameter of the indent made by the sphere. Table 24.3 shows the results for four experiments.

Table 24.3

Height of sphere above clay (cm)	Diameter of indent (cm)
5	1.0
20	1.7
50	2.2
70	2.5

4 How do you think the pupils measured the diameter of the indent?

5 Plot a graph of their results.

6 How do these results compare with the results of the first experiment?

7 Suggest a reason for any differences you describe.

8 Can the graph be used to predict indentations produced by falls from any height greater than 70 cm? Explain your answer.

25 Electricity

Circuits

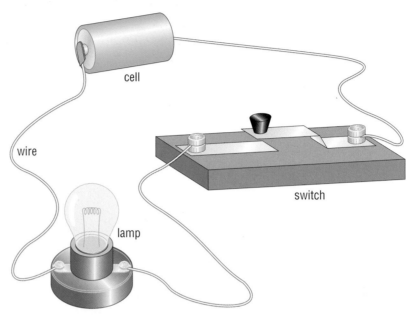

Figure 25.1 A simple electric circuit.

If you set up this equipment and close the switch, the lamp comes on. The wires of the circuit are composed of atoms that are held tightly together but around them are many electrons that are free to move. The metal filament in the lamp and the metal parts of the switch also have free electrons. When the switch is closed, the wires on either side of the switch are linked by metal contacts and a path is made along which the electrons can flow. When you open the switch, the lamp goes out. The path is broken and the electrons cannot flow.

The energy to move the electrons comes from the cell. The chemical reactions that take place in the cell make the electrons leave the cell at the negative terminal when the circuit is completed. They push their way into the wire and move the other electrons along, creating a flow or current of electricity. At the positive terminal electrons are drawn back inside the cell. The wire in the lamp filament is more resistant to the flow of electrons than the other wires in the circuit. As the current moves through the filament some of its electrical energy is transferred to heat energy and light energy.

In time the chemicals which take part in the reaction inside the cell are used up. They can no longer release energy to make the electrons move and the current stops. The number of electrons in the circuit does not change; it is the chemical energy released by the cell that changes.

When Benjamin Franklin (1706–1790) described substances as having positive or negative electric charges, he thought that electricity flowed from a positively charged substance to a negatively charged one. His idea was taken up by other scientists and is still used today. It is known as the conventional current direction.

1 Describe the path of an electron round the circuit in Figure 25.1 when the switch is pressed down.

2 a) How does the wire in the filament behave differently to other wires in the circuit when the current flows?

   b) What property of the wire accounts for this difference?

When circuits are drawn, symbols are used for the parts or components. The use of symbols instead of drawings makes diagrams of circuits quicker to make, and the connections between the components are easier to see. The symbols have been standardised like the SI units described on page 218 and are recognised by scientists throughout the world. The circuit in Figure 25.1 is shown as a circuit diagram using symbols in Figure 25.2a. The components for the circuit are the wires, cell, lamp and switch (Figure 25.2b).

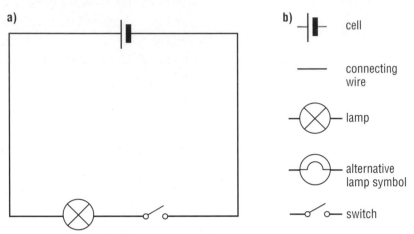

Figure 25.2 **a)** A circuit diagram and **b)** the symbols used.

In everyday life, cells are almost always called batteries but this is scientifically incorrect. In science a battery is made of two or more cells joined together. The symbols for a battery of two cells, three cells or more cells are shown in Figure 25.3a–c.

The lining up of cells next to each other, in a row, end to end as shown in Figure 25.3, is described as arranging them in series. Other electrical components can also be arranged in series, for example, lamps may be arranged in series as Figures 25.5c and 25.6a–e on the next page show.

The voltage of a cell

The ability of the cell to drive a current is measured by its voltage. This is indicated by a figure on the side of the cell with the letter V after it. The volt, symbol V, is the unit used to measure the difference in electrostatic potential energy, usually just referred to as 'potential', between two points. The voltage written on the side of the cell refers to the difference in potential between its positive and negative terminals. It is a measure of the electrical energy that the cell can give to the electrons in a circuit.

When cells are arranged in series with the positive terminal of one cell connected to the negative terminal of the next cell, the current-driving ability of the combined battery of cells can be calculated by adding their voltages. For example, two 1.5 V cells in series produce a voltage of 3 V. The two cells together give the electrons in the circuit twice as much electrical energy as each one would provide separately.

Figure 25.3 **a)** Two cells, **b)** three cells and **c)** any number of cells in series.

a)

b)

c)

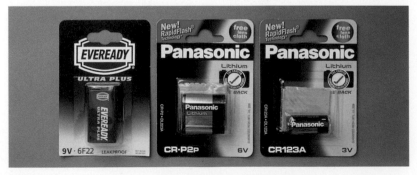

Figure 25.4 The voltage is clearly displayed on the packaging of cells and batteries.

4 A wire carrying a current of electricity can be described as being similar to a stream carrying a current of water. In what ways are the wire and the stream similar?

Resistance

The material through which a current flows offers some resistance to the moving electrons. A material with a high resistance only allows a small current to pass through it when a certain voltage is applied. A material with a low resistance allows a larger current to pass through it for the same applied voltage.

The wires connecting the components in a circuit have a low resistance while the wires in the filaments of the lamps have a high resistance. When the lamps are connected in series their resistances combine in the same way as the voltages of the cells in series – they add. They therefore offer a greater resistance to the current than each lamp would separately.

Lamps and current size

The size of the current flowing though a circuit can be estimated by looking at the brightness of the lamps in the circuit. A lamp shines with normal brightness when it is connected to one cell as shown in Figure 25.5a. The lamp shines more brightly than normal when it is in a circuit with two cells (Figure 25.5b) and shines less brightly when it is in a circuit with one cell and another lamp as shown in Figure 25.5c.

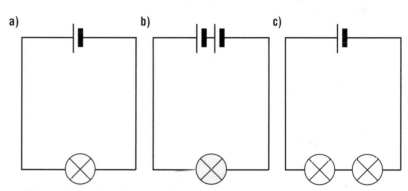

Figure 25.5   Three arrangements of cells and lamps in series.

5 Predict the brightness of the lamps in the circuits in Figure 25.6 compared with that of a single lamp in a circuit with one cell. Use one of the following words in each case:
very dim, dimmer, normal, brighter, very bright.
(All the lamps are identical and all the cells have the same voltage.)

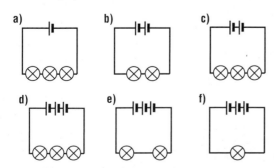

Figure 25.6

6 Compare the circuit in Figure 25.7 with the one in Figure 25.6b. Do you think the lamps will glow with the same brightness? Explain your answer.

Figure 25.7

7 How do you think the brightness of two lamps arranged in parallel compares with the brightness of two lamps arranged in series (both arrangements having one cell)?

8 If current flows through two lamps arranged
   a) in series,
   b) in parallel,
   and the filament of one lamp breaks, what happens to the other lamp? Explain your answers.

9 How many electrons are flowing per second past a point in a circuit in which there is a current of
   a) 0.5 amps,
   b) 5 amps,
   c) 30 amps?

10 Towards which terminal of the power supply should the negative or black terminal of an ammeter be connected?

Parallel circuits

Lamps can be arranged in a circuit 'side by side' rather than end to end. This kind of circuit is called a parallel circuit (Figure 25.8). The resistances of the lamps do not combine to oppose the flow of current in the same way as they do in a series circuit. Each lamp receives the same flow of electrons as it would if it were on its own in a circuit with the cell.

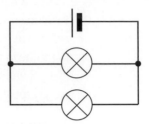

Figure 25.8 Two lamps in parallel.

Measuring current

The rate at which electrons flow through a wire is measured in units called amperes. This word is usually shortened to amps and the symbol for it is A. One amp is equal to the flow of 6 million, million, million electrons passing any given point in the wire in a second!

The current flow in a circuit is measured using an instrument called an ammeter. This is a device which has a coil of wire set between the north and south poles of a magnet. The coil has a pointer attached to it and it turns when a current passes through it. The amount by which the coil turns depends on the size of the current and is shown by the movement of the pointer across the scale.

When an ammeter is used it is connected into a circuit with its positive or red terminal connected to a wire that leads towards the positive terminal of the cell, battery or power pack. It is always connected in series with the component through which the current flow is to be measured (Figure 25.9). Ammeters usually have a very low resistance so that the current passes through them without affecting the rest of the circuit.

When an ammeter is to be used to measure the current flowing through a series circuit such as that shown in Figure 25.10a, the ammeter is placed at a position such as A or B.

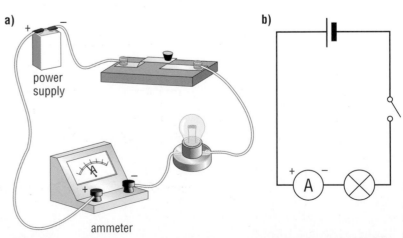

Figure 25.9   a) An ammeter connected in a circuit and **b)** the circuit diagram showing its symbol.

When an ammeter is to be used to measure the current flowing through a parallel circuit such as that shown in Figure 25.10b, the ammeter should be placed at positions A, B and C in turn.

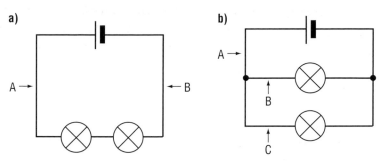

Figure 25.10　Measuring current in **a)** series and **b)** parallel circuits.

Switches

At the voltages used in simple electrical circuits the air does not conduct electricity, so if there is a gap in the circuit a spark does not cross it. The current simply stops flowing and devices in the circuit, such as lamps and motors, stop working.

As you know, the stopping and starting of current flow in a circuit can be controlled by a switch. When a switch is switched on, or closed, a gap in the circuit is bridged by a conducting material through which the current flows. When the switch is switched off, or opened, the conducting material is moved so that a gap is formed and the current stops flowing.

Types of switch

The on/off switch and the push button switch are well known. Their symbols are shown in Figure 25.11.

The on/off switch is also known as the single-pole single-throw switch or SPST switch. The part that moves to close the gap is called the pole, and the action of moving the pole so it makes contact to complete the circuit is called the throw.

Some switches are called single-pole double-throw switches or SPDT switches. They can close a gap to complete a circuit in two ways, as Figure 25.12 shows.

Switches that rely on electromagnetism are described in Chapter 29.

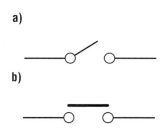

Figure 25.11　Symbols for **a)** an on/off switch and **b)** a push button switch.

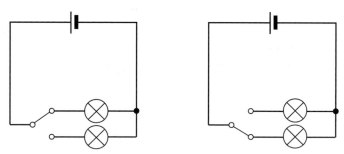

Figure 25.12　Circuit with a single-pole double-throw switch.

Controlling components with switches

Two switches may be used in a circuit to control the operation of an electrical device. They may be arranged in series or in parallel.

When the switches are arranged in series they make an AND circuit. This means that you must close switch 1 AND switch 2 for a current to flow (Figure 25.13).

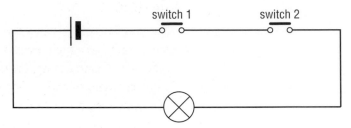

Figure 25.13   AND circuit.

When the switches are arranged in parallel they make an OR circuit. This means that you must close switch 1 OR switch 2 OR both for the current to flow (Figure 25.14).

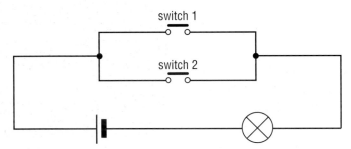

Figure 25.14   OR circuit.

The position of the switches in a circuit and the action of the electrical component, such as a motor or an LED (see page 253), can be given in a truth table. A truth table for the AND circuit is shown in Table 25.1.

Table 25.1   Truth table for the AND circuit.

Switch 1	Switch 2	Component
open	open	off
open	closed	off
closed	open	off
closed	closed	on

Truth tables are very useful for describing the positions of switches and the action of components in the complicated circuits in microprocessors.

An AND circuit is used in a pedestrian crossing control. The length of time the traffic lights show is controlled by a timer and a switch. When the green traffic light has shown for a certain length of time, the

11 Construct a truth table for the OR circuit of Figure 25.14, using Table 25.1 as a guide.

12 a) Which circuit would you use if two switches are required to be closed in sequence for an electric motor to open a bank vault door?

   b) Draw a circuit diagram (see Figure 29.14 on page 294 for the motor symbol).

13 a) How could you use an OR circuit to protect two windows with a burglar alarm?

   b) Draw a circuit diagram (see page 253 for the buzzer symbol).

timer closes one switch. If you then press the pedestrian button, a second switch is closed and the circuit to make the red traffic light come on is activated.

Street lighting may be controlled by a switch that can be operated manually or by a light dependent resistor, using an OR circuit.

Other circuit components
Resistors

The property of resistance was discussed on page 247. If a short piece of wire which has a high resistance is included in a simple circuit with a cell, a switch and a lamp, the lamp will shine less brightly than before. If a longer length of high resistance wire is included in the circuit the light will shine even more dimly.

A component that is designed to introduce a particular resistance into a circuit is called a resistor.

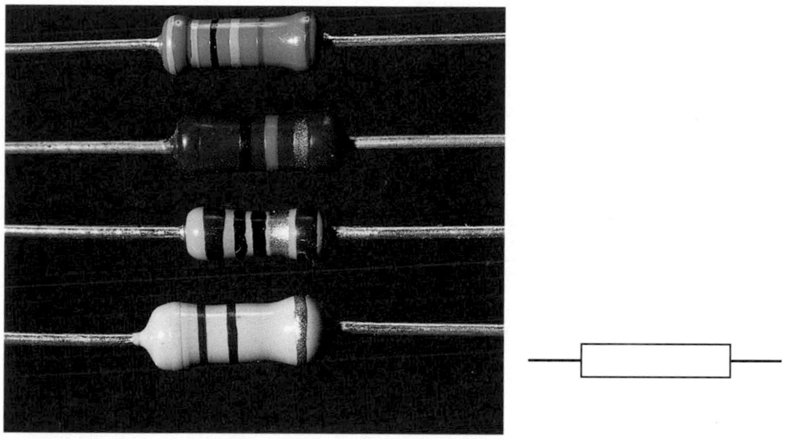

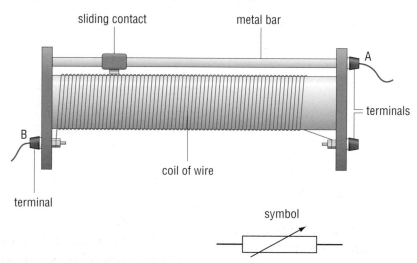

Figure 25.15   Four resistors and the symbol for a resistor.

A variable resistor can be made in which a contact moves along the surface of a resistance wire and brings different lengths of the wire into the circuit. This device is sometimes called a rheostat. In order to make it more compact, the length of the wire is wound in a coil and the contact is made to move freely across the top of the coil.

Figure 25.16   A variable resistor and its symbol.

14 How does the length of a high resistance wire affect the flow of current through the circuit?

15 In Figure 25.16, which way should the contact be moved to
a) increase,
b) decrease
the resistance in the part of the wire included in the circuit between A and B?

16 Figure 25.17 shows a variable resistor in a dimmer switch. How would you turn the switch to make the lights
a) brighter,
b) dimmer?
Explain your answer.

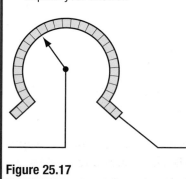

Figure 25.17

In Figure 25.16 the current passes through terminal A, along the bar, through the sliding contact and coil of wire to terminal B. When the contact is placed on the far left the current passes through only a few coils of the wire. As the contact is moved to the right the current flows through more of the wire and encounters greater resistance. When the contact is moved from the right to the left the current flows through fewer coils of the wire and encounters less resistance.

The LDR

An LDR is a light dependent resistor. It is made from two pieces of metal which are joined together by a semiconductor. A semiconductor is a material that has just a few electrons which can move freely. Silicon and germanium are two elements that are widely used as semiconductors.

When the LDR receives light energy, more electrons are released in the semiconductor to move freely through it, and the resistance of the LDR becomes lower. When the amount of light shining on it is reduced, fewer electrons can flow and the resistance increases.

17 Some bedside clocks have a display which glows dimly when the room is dark yet shines brightly if the room is light. How could an LDR be responsible? How may the LDR make it easier for you to get to sleep?

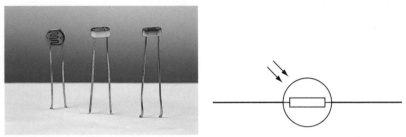

Figure 25.18   Three LDRs and the symbol for an LDR.

Diodes

A diode is a device made from a semiconducting material which has been given two impurities, phosphorus and boron. This allows the current to pass through the diode in one direction only. Diodes are used to control the direction of the flow of a current through complicated circuits, such as those used in a radio, which have components in series and in parallel.

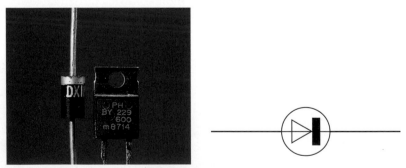

Figure 25.19   Two diodes and the symbol for a diode.

Diodes have a band marked at one end. When the diode is connected into a circuit the end with the band on must be connected to a wire coming from the negative terminal of the cell or battery for the current to flow. When a diode symbol is drawn in a circuit diagram the symbol should be drawn with the straight line facing the negative terminal of the source of the current.

18 Compare the action of a resistor and a diode.

The LED

An LED is a light emitting diode. In simple circuits we often use a lamp to show that a current is flowing. In electronic circuits an LED performs the same task more efficiently. An LED is a semiconductor diode, allowing a current to flow in only one direction through it, and it produces light. It does not produce light in the same way as a filament in a lamp. It is made from semiconductors such as gallium arsenide and gallium phosphide, and when a current passes through these materials their electrons move between distinct energy levels and release some of their energy as light. An LED can emit red, yellow, blue or green light. The colour emitted depends on the semiconductor materials used to make the LED.

19 How is an LED
  a) similar to, and
  b) different from a lamp?

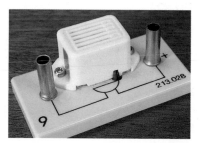

Figure 25.20 Three LEDs and the symbol for an LED.

Buzzers

A buzzer is an electrical device in which one part vibrates strongly when a current of electricity passes through it. The vibrations produce the sound.

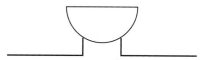

Figure 25.21 A buzzer and its symbol.

Fuses and circuit breakers

When a current flows through a circuit some energy is lost as heat (thermal) energy. If the size of the current increases, the amount of heat (thermal) energy released also increases. A fuse is a device that contains a wire which melts when the current flowing through it reaches a certain value. When the wire melts, it breaks and so also breaks the circuit and stops the current flowing.

Electrical appliances are designed to work when a current of a certain size flows through them. If the current is too large, the appliance may be damaged. An unusually large current can occur in a household circuit in two ways. It may occur when the insulation in a cable is worn and the wires in the cable touch each other. This causes a short circuit. It may also occur if too many appliances are plugged into one socket.

Fuses are used to stop the flow of a current when it becomes too large for the circuit. They may be present in plugs (see Figure 25.22c) and/or in the appliances themselves. In the past each circuit in a home was protected by a fuse in a consumer unit or 'fuse box'. Today circuit breakers are used instead of fuses in consumer units. A circuit breaker is a switch that is sensitive to the size of the current flowing through it. If the current is too large, the switch opens and breaks the circuit. The switch can be closed and the circuit used again once the cause of the problem has been identified and corrected.

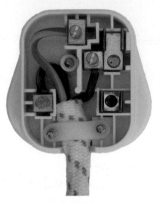

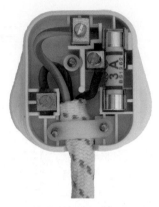

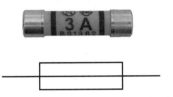

a) A three amp fuse and the circuit symbol for a fuse.

Figure 25.22

b) The inside of a plug with no fuse.

c) The inside of a plug with a fuse in place.

Generating electricity

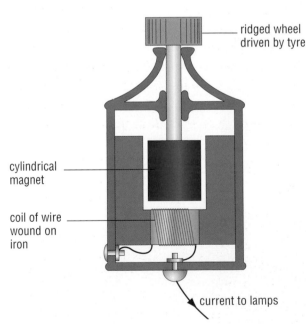

Figure 25.23 Inside a dynamo.

ridged wheel driven by tyre

cylindrical magnet

coil of wire wound on iron

current to lamps

Michael Faraday (1791–1867), an English physicist, discovered that an electric current could be made to flow in a wire if the wire was made to move through a magnetic field. (You can read more about how electricity and magnetism are linked in Chapter 29.) This principle is used to generate electricity in a bicycle dynamo and in a power station generator.

The bicycle dynamo

A bicycle dynamo is an electrical device which is clamped onto the frame of a bicycle close to a tyre. It has a wheel on top which can be made to touch the tyre. The inside of a dynamo is shown in Figure 25.23.

When the dynamo wheel is in contact with the tyre it rotates as the bicycle wheel turns. Inside the dynamo the magnet turns and its field sweeps through the wires, generating an electric current which lights the bicycle's lamps.

The power station generator

Inside a power station there is a generator consisting of a huge electromagnet (see Chapter 29) surrounded by coils of wire. The electromagnet is attached to a shaft to which turbine blades are attached (Figure 25.24). When the turbines are made to spin, the electromagnet also spins, generating a current of electricity in the surrounding coils of wire.

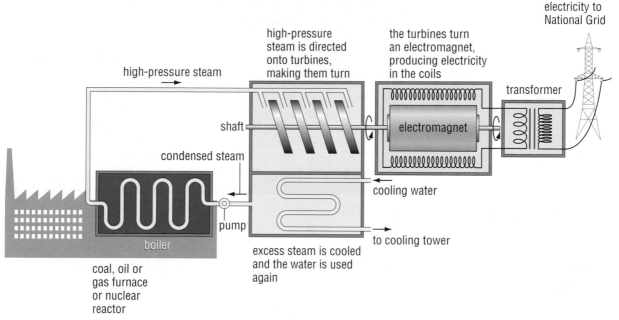

Figure 24.24 The parts of a power station.

In about two-thirds of the world's power stations water is heated to make steam. This takes place in a boiler. The energy that the water molecules receive increases their kinetic energy so much that they move apart from each other to form a gas – steam. The steam expands rapidly and exerts a force which drives it from the boiler to the turbine blades. Here as much as possible of the kinetic energy of the steam is passed to the turbine blades as the steam pushes past them, making the blades spin on the central shaft.

The generator's electromagnet is connected to the end of the shaft. As it spins using kinetic energy from the turbine blades it generates a current of electricity in the coils of wires surrounding it. The electricity flows away from the power station to towns and cities in overhead power lines or underground cables.

Most power stations use fossil fuels to produce the heat to make steam. In some gas-powered stations the combustion gases are used as well as steam to turn the turbines.

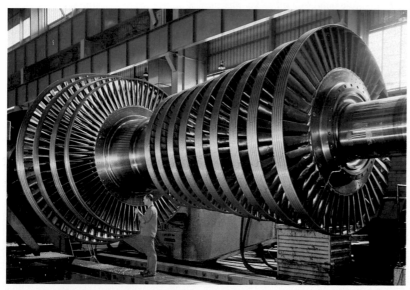

Figure 25.25 Turbine assembly.

Nuclear power stations use heat produced in a nuclear reactor to make steam. Geothermal power stations use the heat from underground rocks to produce steam. In other power stations steam is not used at all. The turbines are made to spin by water or wind (see Figures 24.21 and 24.22, page 239).

Figure 25.26 A gas turbine power station.

Summary

- A closed or complete circuit is needed for an electric current to flow (*see page 245*).
- The ability of a cell to drive a current round a circuit is indicated by its voltage (*see page 246*).
- The material through which a current flows offers resistance to the current (*see page 247*).
- Lamps and other components may be arranged in series or in parallel (*see page 247*).
- An ammeter measures the rate of flow of the current (*see page 248*).
- The starting and stopping of current flow in a circuit is controlled by one or more switches (*see page 249*).
- Variable resistors and light dependent resistors can be used to control current flow (*see page 250*).
- A diode is used to control the direction of current flow through a circuit (*see page 252*).
- A buzzer vibrates when a current passes through it (*see page 253*).
- Fuses and circuit breakers are used to protect circuits from carrying a dangerously large current (*see page 253*).
- Electricity can be generated by magnetism (*see page 254*).
- Electricity is generated on a large scale in power stations (*see page 255*).

End of chapter question

1 A circuit contains a cell, an LDR, an LED and a switch. When the switch is closed and the circuit is left in daylight the LED glows, but when the closed circuit is left in the dark the LED no longer glows. Explain what is happening in the circuit in both the light and the dark.

26 Forces

You cannot see a force but you can see what it does. You can also feel the effect of a force on your body. A force is a push or a pull.

Figure 26.1 Forces act in many ways.

1 Describe the pushing and pulling forces shown in Figure 26.1a–h.

What forces do

- A force can make an object move. For example, if you throw a netball your muscles exert a pushing force on the ball and it moves through the air when you let it go.
- A force can make a moving object stop. For example, a goalkeeper moves into the path of a moving ball to exert a pushing force on the ball to stop it.

- A force can change the speed of a moving object. For example, a hockey player uses a hockey stick to push a slow moving ball to send it shooting past a defender.
- A force can change the direction of a moving object. For example, a batsman can change the direction of a cricket ball moving towards the wicket by deflecting it so that it moves away from the wicket towards the boundary.
- A force can change the shape of an object. For example, when a racket strikes a tennis ball, part of the ball is flattened before the ball leaves the racket.

Figure 26.2 A force makes the ball move.

Figure 26.3 A force makes the ball stop.

Figure 26.4 A force increases the speed of the ball.

Figure 26.5 A force changes the direction of the ball.

Figure 26.6 A force changes the shape of the ball.

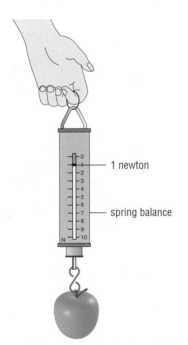

1 newton

spring balance

Figure 26.7 The apple pulls on the spring balance with a force equal to its weight.

How to measure a force

A force can be measured with a newton spring balance. The SI unit for measuring force is the newton (symbol N). This force is quite small and is equal to the gravitational force on (the weight of) an average-sized apple, or the pulling force needed to peel a banana!

Different types of forces

There are two main types of forces: contact forces and non-contact forces. A contact force occurs when the object or material exerting the force touches the object or material on which the force acts. A non-contact force occurs when the objects or materials do not touch each other.

Figure 26.8 The car behind exerts an impact force on the car in front.

Contact forces

All the situations described so far have been examples of contact forces in action. Some more examples follow.

Impact force

When a moving object collides with a stationary object an impact force is exerted by one object on the other. The size of the force may be large such as when a hammer hits a nail or it may be very tiny such as when a molecule of gas in the air strikes the skin.

Strain force

When some materials are squashed, stretched, twisted or bent they exert a force which acts in the opposite direction to the force acting on them. These materials are called elastic materials and the force they exert when they are deformed is called a strain force. When the force applied to the material is removed the strain force exerted by the material restores the deformed material to its original shape. For example, the strain force in the squashed tennis ball in Figure 26.6 returns the ball to its original shape when the ball has left the racket.

Tension is a strain force that is exerted by a stretched spring, rope or string. At each end the tension force acts in the opposite direction to the pulling force.

A force is shown in a diagram as an arrow pointing in the direction of the push or the pull.

> **2 a)** i)   What do you feel if you hook the two ends of an elastic band over your index fingers and slowly move your hands apart?
> ii) What happens to the elastic band when you bring your hands together again?
> iii) What happens if one end of a stretched elastic band is released?
> **b)** Describe how the strain force changes in each part of question a).

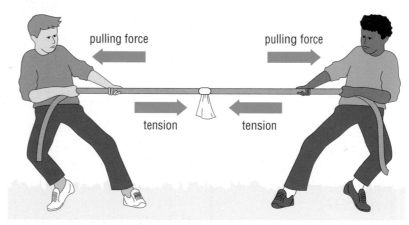

Figure 26.9 Tension is a strain force which acts against the force applied.

Friction

This contact force occurs between two objects when there is a push or a pull on one of the objects which could make it move over the surface of the other object. Friction acts to oppose that movement.

As the push or pull on the object increases, the force of friction between the surfaces of the objects also increases. This force matches the strength of the push or the pull up to a certain value and so, below this value, the object does not move. The friction which exists between the two objects when there is no movement is called static friction.

If the strength of the push or pull on the object is increased beyond this value the object will start to slide. There is still a frictional force between the two surfaces, acting on each surface in the opposite direction to the direction of its movement. This frictional force is called sliding friction. The strength of this force is less than the maximum value of the static frictional force.

3 Imagine you are asked to push a heavy box across the floor. At first you need to push very hard but once the box has started to move you can push less strongly yet still keep it moving. Why is this?

4 When you take a step forwards you push backwards on the ground with your foot.

Figure 26.11

Make a sketch of Figure 26.11 and draw in an arrow to show the frictional force that stops your foot slipping.

5 Figure 26.12 shows a wheel turning.

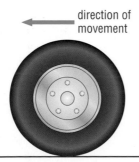

direction of movement

Figure 26.12

Make a sketch of the wheel. Draw and label

a) the force exerted by the wheel pushing backwards on the road and

b) the force of friction preventing the wheel slipping.

Figure 26.10 Friction between the log and the ground opposes the pulling force of the horse.

A closer look at friction

The surfaces of objects in contact are not completely smooth. Under a microscope it can be seen that they have tiny projections with hollows between them (Figure 26.13).

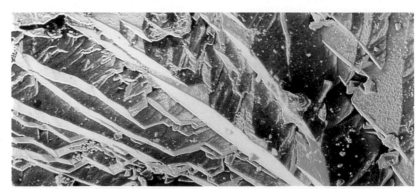

Figure 26.13 A metal surface that appears smooth to the naked eye has projections which can be seen when it is magnified 180 times.

Where the projections from the surface of one object meet the projections from the surface of the other, the materials in the projections stick. These connections between the surfaces produce the force of friction between the objects.

Reducing friction

If a liquid is placed between the two surfaces the projections are forced apart a little and the number of connections is reduced, which in turn reduces the force of friction. This can cause problems or it can be

6 Why does oiling the axles of a bicycle make the bicycle move more easily?

helpful. For example, water running between the surface of a tyre and the road reduces the friction between them and increases the chance of skidding. However, oil between the moving metal parts of an engine and the parts in the bearings reduces friction and also reduces wear on the metal parts.

Increasing friction

The friction between two surfaces can be increased by pressing the surfaces together more strongly. This makes the projections press against each other more strongly and increases the size and number of connections between the surfaces.

When brakes are applied on a bicycle or car the brake pads press against a moving part of the wheel and the force of friction increases. This opposes the rotation of the wheel and slows down the bicycle or car until it stops.

The tread on a car tyre is designed to move water out of the way as the tyre rolls over a wet road, reducing the risk of skidding. Racing cars have smooth tyres that are ideal for a dry track. If it rains they slide and skid all over the place and the tyres need to be changed.

Figure 26.14 Tyres with treads are designed so that water squirts out from between the treads.

Friction and road safety

When a driver in a moving car sees a hazard ahead the car travels a certain distance before the driver reacts and applies the brakes. The distance travelled by the car in this time is called the thinking distance. This is followed by the braking distance, which is the distance covered by the car after the brakes are applied and before the car stops.

Table 26.1 shows the thinking and braking distances that will bring a car with good brakes to a halt on a dry road.

7 a) What happens at the beginning of the time during which a car covers the thinking distance?
 b) What happens at the end of the time during which the car covers the thinking distance?
8 What may affect the thinking time of the driver? How would the thinking distance of the car be affected? Explain your answer.
9 What, other than speed, may affect the braking distance of the car? Explain your answer.
10 A car is travelling along a road at 80 km/h when a tree falls across the road 54 metres away. What would probably happen and why?

Table 26.1

Speed	Thinking distance (m)	Braking distance (m)	Total stopping distance (m)
48 km/h (30 mph)	9	14	23
80 km/h (50 mph)	15	38	53
112 km/h (70 mph)	21	75	96

For discussion

How safe is a) driving close to the car in front, b) driving fast on winding country roads with high hedges? Explain your answers to each part of the question.

'It's the driver that's dangerous, not the car.'

Assess the usefulness of this slogan for a road safety campaign.

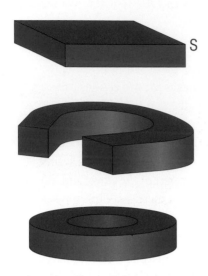

N = north pole
S = south pole

Figure 26.15 Bar, horseshoe and ring magnets.

11 If you had a magnet with its north and south poles marked on it and a magnet without its poles marked, how could you identify the poles of the unmarked magnet? Explain your answer.

12 Compare magnetic and electrostatic forces. In what ways are they
  a) similar,
  b) different?

Non-contact forces

These include magnetic forces, electrostatic forces and gravitational forces. They all exert their force without having to touch the object.

Magnetic force

A magnet has a north-seeking pole and a south-seeking pole. These are usually known as the north pole and the south pole. If you pick up two magnets and bring either their north poles or their south poles together you will feel a force pushing your hands apart as the two similar poles repel each other. You will feel your hands being pushed away even though the magnets are not touching. The strength of the push increases as you bring the two similar poles closer together.

If you bring the north pole of one magnet towards the south pole of another magnet you will feel your hands being pulled together as the different poles attract each other. The strength of this pull increases as the poles get closer together.

A magnet can also exert a non-contact force on objects made of iron, steel, cobalt or nickel. Either pole of the magnet exerts a pulling force on these magnetic materials. The strength of the force increases as the magnet and the magnetic material are brought closer together.

Figure 26.16 This 'Maglev' train is supported above its track by strong magnetic forces. It travels quietly on a 'cushion' of air which eliminates friction between the train and the tracks.

Electrostatic force

If certain electrical insulator materials are rubbed an electrostatic charge develops on them. There are two kinds of charge: positive charge and negative charge. The forces between the charges can be investigated by suspending a plastic rod so that it can swing freely (Figure 26.17), giving the rod an electrostatic charge then bringing rods with different charges close to it. If the suspended rod has a positive charge it will move away from a plastic rod which also has a positive charge, as the similar charges repel each other. If a rod with a negative charge is brought near to the positively charged suspended rod, the rod swings towards it as the opposite charges attract each other. The strength of the force between electrostatic charges increases as the rods are brought closer together.

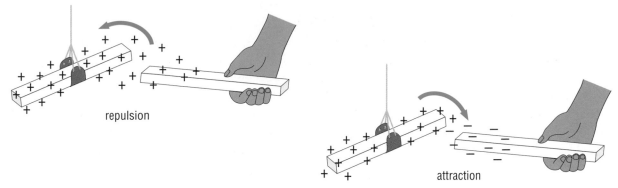

repulsion

attraction

Figure 26.17 Investigating electrostatic force with charged plastic rods.

Gravitational force

There is a force between any two masses in the Universe. The masses may be small such as those of an ant and a pebble or they may be very large such as those of the Sun and the Earth. The force that exists between any two masses because of their mass is called the gravitational force. The force acting between small masses is too weak to have any noticeable effect on them but the gravitational force between two large masses such as the Sun and the Earth is large enough to be very important. It is the gravitational force between the Sun and all the planets in the Solar System that holds the planets in their orbits. The gravitational force between an object on the Earth and the Earth itself pulls the object down towards the centre of the Earth and is called weight of the object.

Mass and weight

● The *mass* of an object is a measure of the amount of matter in it.
● The *weight* of an object is the pull of the Earth's gravity on the object. For example, an object may have a mass of 1 kg. The pull of the Earth's gravity on 1 kg is a force of almost 10 newtons (actually 9.8 N but it is often rounded up to make the calculations easier). The weight of the 1 kg mass is therefore 10 N.

The region in which a force acts is called a field. There is a gravitational field around the Earth. The gravitational field strength is calculated by the equation:

$$\text{gravitational field strength} = \frac{\text{weight}}{\text{mass}}$$

At the Earth's surface we have seen that the pull on a mass of 1 kg is 10 N so the gravitational field strength is 10 N/kg.

The above equation can be rearranged to calculate the weight of an object, given its mass, in a known gravitational field:

$$\text{weight} = \text{mass} \times \text{gravitational field strength}$$

The gravitational field strength on the surface of Mars is one-third of the gravitational field strength on the surface of the Earth. This means that a 1 kg object that is part of a space probe would have a weight

13 What is the weight of the following masses on Earth:
a) 2 kg,
b) 3.5 kg,
c) 5.25 kg?

14 What is the weight of a 6 kg object on the surface of Mars?

15 It is planned to bring samples of Mars rock back to the Earth. If 50 kg samples were collected by a space probe robot, what would be the weight of the rocks on
a) Mars,
b) Earth?

16 The Moon's gravitational field strength is one-sixth that of the Earth. What would be the weight of a 1 kg object on the Moon?

17 A sample of Moon rock weighed 30 N on the Moon.
   a) What would be its weight when it was brought to the Earth?
   b) What is the mass of the sample?

18 A spring is 6 cm long when it is unstretched but is stretched to 9 cm when a mass is hung from it. What is the extension of the spring?

19 An unstretched spring is 6 cm long but it becomes 7 cm long when a 100 g mass is hung from it. The spring becomes 8 cm long when a 200 g mass is hung from it.
   a) What is the extension for each mass?
   b) What extension do you predict when masses of i) 300 g and ii) 350 g are hung from it in turn? Can you be sure that the extension values you predict will in fact occur? (Hint: think about the elastic limit.)

of 10 N when it was on Earth but a weight of only 3.3 N on the surface of Mars.

The mass of an object remains the same wherever it goes in the Universe but its weight changes according to the gravitational force that is acting upon it.

How springs stretch

Robert Hooke (1635–1703) investigated the way in which springs stretched when masses were attached to them. He first hung up a spring and measured its length without any mass attached to it. He then hung a mass on the bottom and measured the new length of the spring. He calculated the extension of the spring by subtracting the original length of the spring from the new length of the spring with the mass attached. Hooke repeated the experiment with different sizes of masses. Each time he found the total extension by subtracting the original length from the new length. He found that as the size of the mass increased the size of the extension increased in proportion: the extension of the spring was proportional to the mass attached to it.

Each time Hooke removed the mass the spring returned to its original length. However, he eventually placed a mass on the spring that stretched the spring so much that it remained slightly stretched when the mass was removed. The spring had gone beyond a point called the elastic limit and was permanently deformed. When a larger mass was then added to the spring it no longer extended in proportion to the mass. The spring beyond its elastic limit was in a state known as plastic deformation (see Figure 26.18).

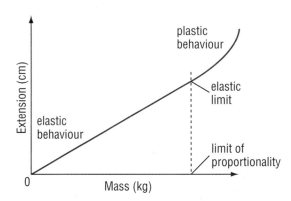

Figure 26.18 How the extension of a spring varies with the mass attached to it.

The newton spring balance

The discovery made by Robert Hooke has led to the development of a force measurer using a spring which is not stretched beyond its elastic limit. This instrument is called a spring balance. The extension of the spring, and hence the reading on a scale, is proportional to the weight of the mass hung from it, or the force with which it is pulled. The scale of the balance is calibrated in newtons so it is sometimes called a newton spring balance or a newtonmeter.

There is a range of spring balances which measure forces of different sizes. For example, a spring balance may measure forces with values in the range 0–10 N, 0–100 N or 0–200 N.

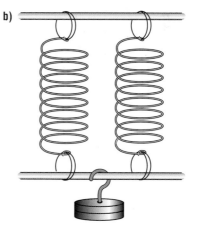

Figure 26.19 Spring balances with different scales.

20 How do you think the spring in a spring balance with a scale of 0–10 N compares with a spring in a spring balance that measures forces up to 500 N?

21 A spring balance without a stop would not give correct readings for the weights of the masses hung from it if large masses were used. Explain the reason for this.

There is a device called a stop on most spring balances. It prevents the spring from stretching beyond its elastic limit.

Combining identical springs

Springs can be combined in two ways. They can be combined in series or in parallel (see Figure 26.20). These terms are taken from the way we describe how components can be combined in electrical circuits (see pages 247 and 248).

22 Two identical springs each extended 2 cm when a mass was added to them separately. They were then connected in series and the same mass hung on the bottom spring.

a) How far did each spring extend when it was arranged in this way?

b) What was the total extension of the two springs together?

23 Two springs each extended 1 cm when a mass was added to them separately.

a) How far did each spring extend when the same mass was added to them when they were connected in parallel?

b) How far did the mass sink as the springs stretched?

a)   b)

Figure 26.20 Identical springs in **a)** series and **b)** parallel.

24 Three similar springs, A, B and C, extended 4 cm when a mass was added to each of them separately. The springs were then arranged as shown in Figure 26.21 and the same mass was attached at point P.

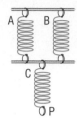

Figure 26.21

a) What was the extension of each of the springs A, B and C?

b) How far altogether did the mass sink when it was attached to point P?

c) How far did the connecting bar move?

Figure 26.22 An action–reaction pair.

25 Describe the forces between you and a chair when you sit on it.

Springs in series

The extension of each spring when a mass is added to the lower one is the same as if each spring were supporting the mass separately. The total extension is therefore double what it would be with just one spring.

Springs in parallel

Each spring, in effect, carries half the mass so it stretches less far. In fact, the extension of each spring when the mass is added to the middle of a light bar connecting them is half the extension it would have if it were supporting the full mass on its own.

Pairs of forces

Action and reaction

A force exerted by one object on another is always accompanied by a force equal to it acting in the opposite direction. For example, if you lean against a wall you exert a contact force that pushes on the wall and the wall exerts a contact force that pushes on you. The forces are equal. The force that you exert on the wall and the force the wall exerts on you are called an action–reaction pair. The action and reaction forces that form the pair act simultaneously: one does not cause the other.

Two model railway trucks each with a spring can be used to demonstrate an action–reaction pair. When the two trucks are pushed together the springs exert a force on each other due to the strain forces which develop in them. The force spring A exerts on spring B is always the same size but opposite in direction to the force exerted by spring B on spring A (Figure 26.23). The action force of A on B pushes B while the reaction force of B on A pushes A the other way. When the two trucks are suddenly released the action–reaction pair between the springs pushes the trucks in opposite directions.

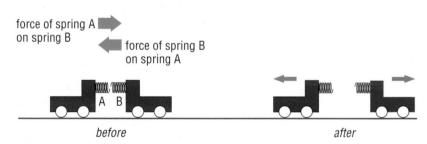

force of spring A on spring B

force of spring B on spring A

before

after

Figure 26.23 Action and reaction forces act in opposite directions.

Balanced forces

A pair of balanced forces is different from the action–reaction pair above, because balanced forces act on one object only. When you stand still you do not rise above the ground or sink into it, because of the two balanced forces acting on you. Your weight acts downwards into the ground because of gravity and the ground exerts a contact force upwards on the soles of your shoes. This contact force is equal to your weight.

A person sitting in a stationary go-kart does not move up or down because the weight of the person and the kart is balanced by the contact force of the ground on the tyres. When the kart is moving in a straight line at a constant speed (Figure 26.24) there is another pair of balanced forces acting on the kart. These forces are the driving force pushing the kart forwards and air resistance pushing backwards on the kart.

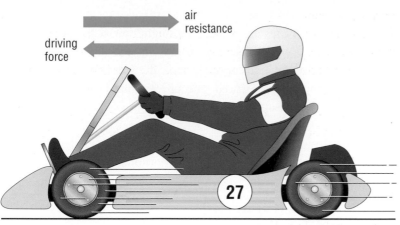

air resistance

driving force

27

Figure 26.24 Balanced forces act on the go-kart when it is moving at a steady speed.

Figure 26.25 The weight of this floating weather station is balanced by the upthrust from the sea water.

Upthrust

When an object is placed in any liquid or gas it pushes some of the liquid or gas out of the way. The liquid or gas pushes back on the object with a force called the upthrust. This force is equal to the weight of the liquid or gas that has been pushed out of the way.

The buoy in Figure 26.25 is carrying a weather station. It floats on the sea surface due to the upthrust acting on it from the sea water.

Unbalanced forces

When the forces on a stationary object are unbalanced the object starts to move. For example, when the driver of a go-kart presses the accelerator pedal on the stationary kart the wheels connected to the engine turn and the frictional force between the ground and the tyres pushes the kart forwards. As the kart moves forwards the air pushes on it (and the driver) with a force called drag or air resistance. To begin with this force is smaller than the frictional force and the kart continues to accelerate forwards (Figure 26.26).

Note that the size of a force on a diagram is indicated by the size of the arrow. A large force is shown by a longer arrow than a small force.

> **26** What are the forces acting on a duck when it floats in water?

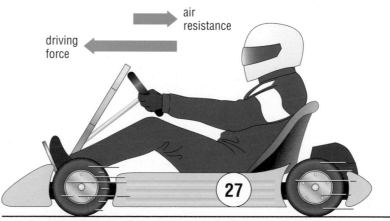

air resistance

driving force

27

Figure 26.26 The forward force is greater than the backward force so the kart speeds up.

27 How does friction help you ride your bike?

28 Draw a submarine sinking in water. Draw and label the forces acting on it and indicate the strength of each one by the size of its arrow.

29 As a space rocket rises from the Earth, how is its weight affected by

a) the Earth's gravitational force,

b) the consumption of fuel?

Figure 26.27 Launch of a space shuttle.

As the kart increases speed the air resistance also increases. Eventually the kart moves at a constant speed in a straight line, as shown in Figure 26.24, with the two horizontal forces balanced.

When the driver takes his or her foot off the accelerator pedal the frictional driving force is reduced. The air resistance is now stronger than the driving force and kart slows down.

The driver can slow down the kart faster by applying the brakes. The brake pads exert a frictional force on the wheels which makes it harder for the wheels to turn. This produces an additional resistance backwards which slows the kart down.

Launching spacecraft

A spacecraft fired directly towards distant parts of the Solar System from the launch pad would have to leave the Earth at the colossal speed of 40 000 km/h (the escape velocity) to escape the pull of the Earth's gravity.

In fact, spacecraft are launched into space in stages by taking them first on a rocket into orbit around the Earth before they begin their trip to other parts of the Solar System and beyond. By launching a spacecraft in this way the escape velocity does not have to be reached and there is a huge saving in fuel.

The rockets that launch spacecraft into orbit still have to be very powerful to raise the weight of the rocket, the spacecraft and the fuel. The reaction force of the hot combustion gases leaving the rocket engines lifts the rocket.

Satellites

Artificial satellites are spacecraft that stay in orbit round the Earth. There are hundreds of them. Some have cameras on board and take pictures of the Earth. These pictures are used for studying and predicting the weather or for monitoring worldwide crop production. Many satellites are communication satellites. They relay telephone messages, television programmes and Internet communications around the Earth.

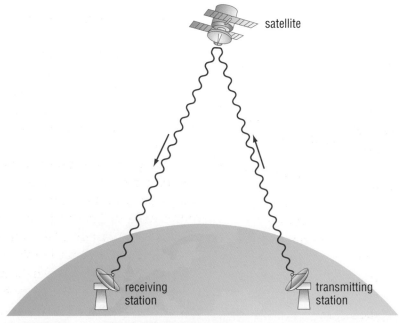

Figure 26.28 A satellite communication link.

Some satellites are used for navigation. The signals they send out are used by the crews of ships and aircraft to keep on course as they travel round the Earth. The Global Positioning System or GPS is a system of 24 satellites which can tell a person his or her position on the planet very accurately (to within a metre). Information from four of the satellites is gathered by a hand-held receiver about the size of a mobile phone.

A few satellites, such as the Hubble space telescope, investigate space. This telescope is not affected by the atmosphere as telescopes on the Earth's surface are, so it can obtain clearer pictures of deep space.

Figure 26.29   The Hubble space telescope being serviced in orbit.

Satellites are launched into orbit by rockets. Once a satellite has reached its orbit, panels of solar cells called a solar array unfold and collect light energy from the Sun (see Figure 24.20 on page 238, and Figure 26.29). The light energy is converted into electrical energy which is used to work the equipment on board. The satellite is controlled from a ground station. Messages are sent on a beam of radio waves to the satellite where a receiver converts them into electric currents. Messages from the satellite are transmitted back to the ground station, also by radio waves. The position of the satellite can be altered by commanding thruster rockets on the side of the satellite to fire.

If the satellite becomes damaged or fails to work, it may be repaired in space by astronauts as in Figure 26.29, or it may have to be brought back to Earth.

Satellites that orbit close to the Earth have to travel faster than those that orbit further away because the Earth's gravity is stronger closer to the planet. A satellite must travel at a speed that will prevent it being pulled down to Earth by the force of gravity. A satellite in a low orbit may therefore pass round the planet several times a day.

At a distance of 35 880 km above the equator a satellite can orbit at the same rate as that at which the Earth turns. This makes the satellite stay over a particular place all the time while the Earth rotates. This orbit is called a geostationary orbit and is used by communication and navigation satellites.

For discussion

What will happen to a satellite if its speed is a) too low for its orbit, b) correct for its orbit, c) too fast for its orbit?

30 What is the advantage of a geostationary orbit rather than a lower orbit for a communication satellite? Look back at Figure 26.28 to help you answer.

Four satellite orbits are shown in Figure 26.30. Satellites that monitor the weather move in a polar orbit. From this path the satellite can take pictures of all parts of the Earth as it rotates below the satellite. Some satellites have eccentric orbits. They monitor conditions in the Earth's magnetic field (see page 290).

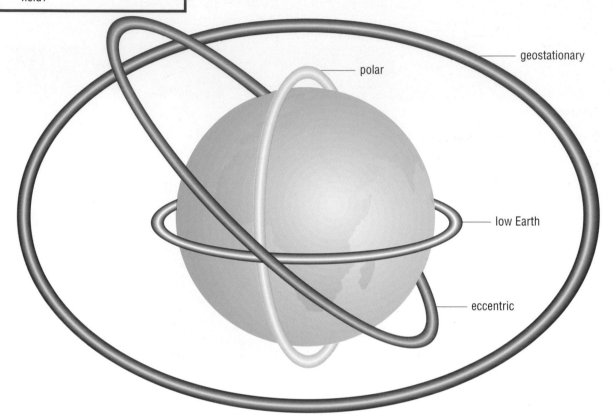

Figure 26.30 Different satellite orbits.

Summary

- A force is a push or a pull (*see page 257*).
- A force can make an object move, stop, change speed or change direction, or can change the object's shape (*see page 257*).
- Examples of contact forces are impact forces, strain forces and friction (*see page 259*).
- Examples of non-contact forces are magnetic forces, electrostatic forces and gravitational forces (*see pages 262–263*).
- The weight of an object is the pull of the Earth's gravity on the object (*see page 263*).
- Springs stretch when forces are applied to them (*see page 264*).
- Forces act in pairs (*see page 266*).
- Forces on an object may be balanced (*see page 266*).
- Forces on an object may be unbalanced (*see page 267*).
- Spacecraft are launched in stages due to the effect of the Earth's gravity on them (*see page 268*).
- Artificial satellites in orbit around the Earth have a wide range of uses (*see page 268*).

End of chapter questions

1 Identify the forces acting in this scene.

Figure 26.31

2 How do artificial satellites help people living on the Earth?

27 The Solar System and beyond

Lights in the sky

The Sun is the only large luminous body in our Solar System. We see the other large objects, such as the Moon, the planets and comets, by the sunlight they reflect to the Earth. As all these objects are relatively close to the Sun and the Earth, they reflect light quite strongly towards the Earth. The other stars are luminous but much further away. Although they generate their own light the beam is very weak by the time it reaches the Earth.

Figure 27.1 The planet Venus shining brightly against a background of stars.

The movement of the gases in the Earth's atmosphere does not significantly affect the strong light beams from the Moon, the planets and comets – they shine steadily in the sky. The weak light beams from the stars, however, are affected and the result is that their light does not shine steadily but appears to flicker or twinkle.

The brightness of a star depends on its size, its temperature and its distance from the Earth. Table 27.1 shows some stars arranged in decreasing order of brightness as seen from Earth.

1 How can you tell a planet from a star in the night sky?

2 a) Look at Table 27.1. Which star is the brighter, Betelgeuse or Spica?

b) Why might you expect Spica to be above Betelgeuse in the table?

c) Betelgeuse is a red giant star. How could this information help you explain its position in the table?

Table 27.1 The features of some bright stars, arranged in decreasing order of brightness.

Star	Temperature (°C)	Colour	Distance from Earth*
Sun	6 000	yellow	8 light minutes
Sirius	11 000	white	8.6 light years
Arcturus	4 000	orange	36 light years
Betelgeuse	3 000	red	520 light years
Spica	25 000	blue	220 light years

*See *Measuring with light*, page 279.

Constellations and planets

The stars make patterns in the sky. These patterns are called constellations. The arrangement of the stars in a constellation is due to their positions in space which in turn are just due to chance. The stars may seem to be grouped together at the same distance from the Earth but they are not. Some stars in a constellation may be many light years closer to the Earth than others.

While the stars appear to be fixed in their positions, the planets do not. Each night a planet is found in a different position from the previous night. The name planet comes from the Greek word for wanderer. The planets wander across the night sky against the background of constellations. This is due to the orbital motion of the Earth and the planets.

Movements in the sky

If you were to watch the sky for a day and a night, you would see the Sun rise towards the east at dawn. It would continue to rise in the sky until midday, then it would sink in the sky and set towards the west. As the sunlight faded the sky would darken and other stars would be seen to cross the sky from east to west before they faded as the sunlight appeared again in the sky. People once believed that these movements of the stars really took place but today we understand that the Sun and the stars do not change position in this way. It is the daily rotation of the Earth that makes them appear to move.

The axis about which the Earth rotates is not perpendicular to the plane of the Earth's orbit. If it were, the Sun would rise to the same height in the sky each day of the year. The axis is at an angle of about 23° to the perpendicular and remains pointing in the same direction throughout the Earth's orbit (see Figure 27.2).

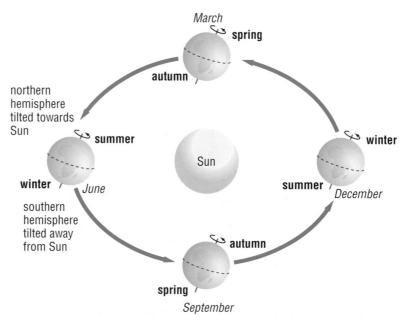

Figure 27.2 The changing seasons in each hemisphere as the Earth progresses in its orbit.

We divide the Earth into two half spheres or hemispheres. They meet at the equator, which is an imaginary line running around the middle of the planet between the poles. The hemispheres are known as the northern hemisphere and the southern hemisphere.

As the Earth moves in its orbit there is a time of year when the northern hemisphere is tilting towards the Sun and the southern hemisphere is tilting away from it. Six months later the northern hemisphere is tilting away from the Sun and the southern hemisphere is tilting towards it.

These changes in the way each hemisphere tilts towards and away from the Sun cause changes in the length of day and night, and in the strength of the sunlight reaching an area of the Earth's surface. This produces the periods of time called seasons.

The east to west path of the Sun across the sky changes with the position of the Earth in its orbit. When a hemisphere is tilting towards the Sun, the path of the Sun is different from the path when the hemisphere is tilting away from the Sun. Sunrise is earlier, the Sun rises higher in the sky at midday and sets later in the evening. Figure 27.3 shows the path of the Sun across the sky when the hemisphere is tilted towards the Sun (mid-summer), away from the Sun (mid-winter) and when it is changing from tilting in one direction to the other (at the spring and autumn equinoxes).

3 What is the position of the Earth when it is summer in
   a) the northern hemisphere,
   b) the southern hemisphere?

4 What is the position of the Earth when the Sun rises to its lowest midday position in the sky in
   a) the northern hemisphere,
   b) the southern hemisphere?

5 What is the position of the Earth when
   a) the day is longer than the night in the southern hemisphere,
   b) the day is shorter than the night in the southern hemisphere,
   c) the day and the night are the same length of time?

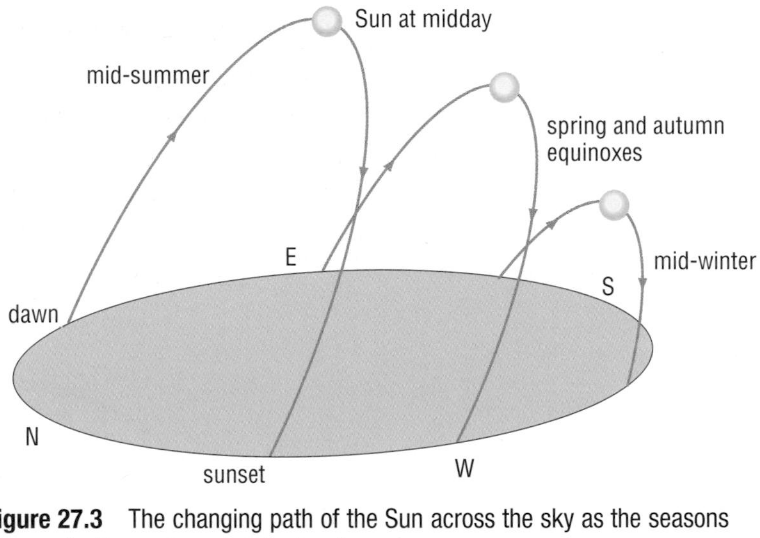

Figure 27.3 The changing path of the Sun across the sky as the seasons change (in the northern hemisphere).

The Moon

The Moon moves round the Earth in about 28 days. Only the side of the Moon's surface that is facing the Sun reflects light, so as its orbit progresses the illuminated part that we can see from Earth changes shape. The different shapes are known as phases of the Moon (Figure 27.4).

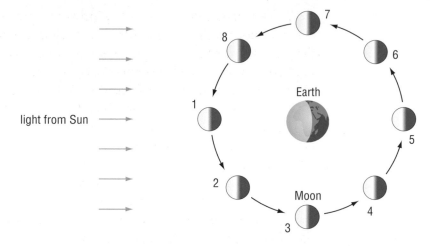

light from Sun →

Earth

Moon

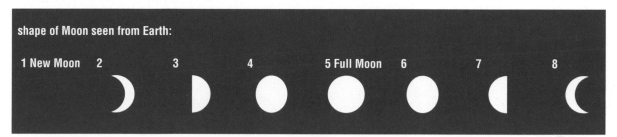

shape of Moon seen from Earth:

1 New Moon 2 3 4 5 Full Moon 6 7 8

Figure 27.4 Phases of the Moon.

Eclipses of the Sun and the Moon

The imaginary surface in which an orbit lies is called the plane of the orbit. Figure 27.5 shows that the plane of the orbit of the Moon is different from the plane of the orbit of the Earth around the Sun. This means that even when the Moon passes between the Sun and the Earth in its orbit there is not always an eclipse of the Sun. In fact, eclipses of the Sun are very rare.

Sun

Moon

Earth

Figure 27.5 Planes of the orbits of the Moon and the Earth.

When the Sun, Moon and Earth do line up exactly a total eclipse of the Sun occurs for viewers on a certain part of the Earth's surface (Figure 27.6, overleaf). The Moon blocks out the light of the Sun.

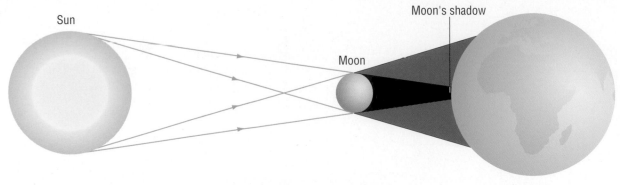

Figure 27.6 An eclipse of the Sun.

Sometimes the Sun, Earth and Moon line up as in Figure 27.7 and an eclipse of the Moon takes place. The Earth blocks out light from the Sun that would normally fall on the Moon at Full Moon phase.

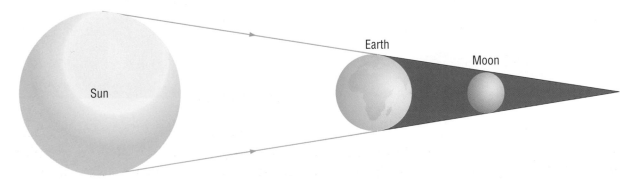

Figure 27.7 An eclipse of the Moon.

Figure 27.8 The Earth's shadow on the Moon during a partial eclipse of the Moon.

> **6** How are an eclipse of the Sun and an eclipse of the Moon
> **a)** similar,
> **b)** different?

> **7** Where did the materials come from to form the Solar System?

The Solar System

It is thought that about 5000 million years ago the Solar System began to form from a huge cloud of gas and stellar material. An exploding star nearby could have caused the cloud to rotate. As the cloud turned it formed a disc. Hydrogen and helium collected at the centre and formed a star, our Sun. The Sun is a middle-sized star called a yellow dwarf star, powered by the fusion of hydrogen into helium. The material moving round the Sun eventually formed the planets and other bodies of our Solar System.

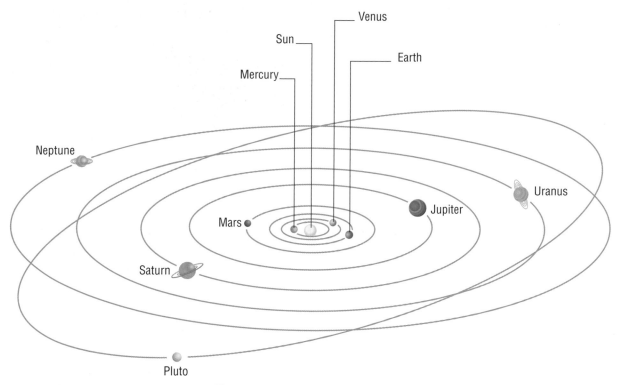

Figure 27.9 The Solar System (not to scale).

For discussion

Living things use the materials in the Earth and the atmosphere to make their bodies. Can they be described as being made from star dust? Explain your answer.

The five planets nearest to the Earth can be clearly seen with the naked eye and were known to the people of the ancient civilisations. The other three planets were discovered by the use of telescopes in the 18th to 20th centuries.

In 2004 a large orbiting object was confirmed to have been found beyond Pluto. It is large enough to be considered by some scientists as a tenth planet (see overleaf) and has been named Sedna.

The planets

Table 27.2 shows some data about the nine known planets.

Table 27.2 Planet data.

Planet	Diameter in km	Mass (Earth = 1)	Distance from Sun in million km (approx.)	Rotation time in			Orbit time in days
				days	hours	minutes	
Mercury	4878	0.056	58	58	15	30	88
Venus	12100	0.82	108	243	0	0	224
Earth	12756	1	150		23	56	365
Mars	6793	0.107	228		24	37	686
Jupiter	142880	318	778		9	50	4332
Saturn	120000	95	1427		10	14	10759
Uranus	50800	14.5	2871		10	49	30707
Neptune	48600	17	4497		15	48	60119
Pluto	2250	0.002	5914	6	9	17	90777

8 Arrange the nine planets in Table 27.2 in order of mass, starting with the most massive planet.

9 Arrange the nine planets in order of diameter, starting with the planet with the largest diameter.

10 Does the diameter of a planet give you an indication of its mass?

11 Does the mass of a planet give you an indication of
   a) its distance from the Sun,
   b) its rotation time?

12 The asteroid belt lies between the orbits of two planets. Which ones? Use Table 27.2 on page 277 to help you.

13 Which asteroids could crash into the Earth?

For discussion

It has been calculated that asteroid 1950DA will hit the Earth in 2880. What should be done?

The masses of the planets are huge numbers. For example, it has been estimated that the mass of the Earth is

$$5\,974\,000\,000\,000\,000\,000\,000\,000\,000 \text{ kg} \quad \text{or} \quad 5.974 \times 10^{27} \text{ kg}$$

The mass of Jupiter by comparison has been estimated as 1.899×10^{30} kg. Using huge numbers like this to make comparisons is difficult. It is easier if the mass of the Earth is taken as 1, as in Table 27.2, and the masses of the other planets are compared with this value. For example, if you look in the table you will see that the mass of Jupiter is 318 times greater than the mass of the Earth.

Sedna – the tenth planet?

In 2004 a large object was confirmed to be orbiting the Sun 3200 million kilometres beyond Pluto. The object has a diameter of 1900 kilometres and is believed to be made of rock and ice. It was discovered by a team of astronomers working in the United States, using data from a space telescope. The object has been officially labelled as 2003VB12 but has also been named after the Inuit goddess of the sea – Sedna. The goddess was believed to live in deep, dark, cold arctic waters, so the name seemed appropriate as the new object is in a very dark, very cold region of the Solar System.

An object has to be a certain size before it can be classed as a planet, and some astronomers consider even Pluto too small to qualify. As Sedna is slightly smaller than Pluto, they do not consider it to be a planet. However, other astronomers think that objects of this size, of the order of 2000 kilometres across, should be classed as a planet.

Asteroids

Asteroids are lumps of rock which move in orbits round the Sun. They range in size from grains of sand to Ceres, the largest asteroid, which is 913 kilometres long. Most of the asteroids move in orbits which are between 300 and 500 million kilometres from the Sun. They form a huge ring of space rubble called the asteroid belt. Some asteroids have orbits further away from the Sun and a few have orbits which take them across the Earth's orbit.

Comets

Comets are bodies in the Solar System that move in orbits round the Sun in the same way as the planets, but their orbits are very elliptical (elongated). Some stretch out into space well beyond Pluto. It is believed they come from a cloud of ice and rock at the very edge of the Solar System. This cloud is called the Oort cloud.

In February 2004 the European Space Agency launched the Rosetta probe. It will move through the Solar System using the 'sling shot method' (using the gravitational effect of planets it passes to redirect it) and should reach its destination, Comet Churumov-Gerasimenko, in November 2014. (How old will you be then?) The probe has two parts. The orbiter will move in a circle around the comet while the lander will settle on the comet surface. Both probes will send detailed information about conditions on and around the comet that will help scientists understand how and when comets formed.

Beyond the Solar System

The Milky Way

The work of many astronomers has shown that the Sun is not the centre of the Universe as people once believed. It is on an arm of a spiral galaxy called the Milky Way (Figure 27.10). This galaxy received its name from the pale white glow it makes across the sky (Figure 27.11).

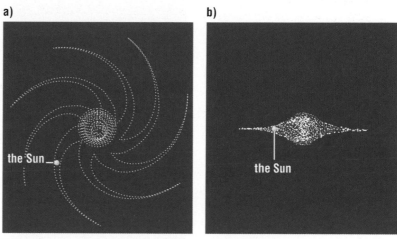

a)

the Sun

b)

the Sun

Figure 27.10 The Milky Way: **a)** top view, **b)** side view.

Figure 27.11 The Milky Way as seen from Earth.

There are about 30 galaxies relatively close to the Milky Way. They form a group of galaxies called the Local Group. Beyond them in all directions are other groups of galaxies at distances of up to thousands of millions of light years away.

For discussion

Some stars are believed to have planets. What do you think are the chances of simple life forms existing on a planet somewhere in the Universe?

How would you rate the chances of discovering the following: a) simple life forms, b) intelligent life forms with a lower technological development than us, c) intelligent life forms with a higher technological development than us? Explain your answers.

How will the vast distances between planets in different parts of the Universe hinder communications if we find intelligent life?

Measuring with light

The vast distance between two objects in space can be measured by the time it takes light to travel between them. For example, the time taken for light to travel between the Sun and the Earth is about eight minutes. The time taken for light to travel between the Sun and Pluto is about five and a half hours. The time for light to travel between two stars is much longer and is measured in light years. A light year is the distance travelled by light in a year. This distance is 9.5 million million kilometres.

The nearest star to the Sun is Proxima Centauri which is 4.3 light years away. This star and the Sun are just two of the 500 000 million stars in the Milky Way galaxy. This is a group of stars which is 100 000 light years across. There are about 100 000 million other galaxies in the Universe. They are great distances from our own. For example, the Andromeda galaxy, which can be seen as a fuzzy patch with the naked eye, is 2.2 million light years away.

Figure 27.12   The Andromeda galaxy seen through a telescope.

Summary

● The Sun and stars are luminous objects (*see page 279*).
● The changes of day and night and the seasons are due to the rotation (spin) of the Earth, the tilt of its spin axis and its orbital path around the Sun (*see pages 273–274*).
● The phases of the Moon are due to the Moon's orbit around the Earth (*see page 274*).
● There can be eclipses of the Sun and the Moon (*see pages 275–276*).
● Our Solar System formed about 5000 million years ago (*see page 276*).
● Asteroids are lumps of rock that move around the Sun (*see page 278*).
● Comets move in very elliptical orbits around the Sun (*see page 278*).
● The Solar System is part of the galaxy called the Milky Way (*see page 279*).

End of chapter questions

Use the information in Table 27.3 opposite to answer the following questions.

1 According to the data in the table, how many moons have been observed in the Solar System? Even more moons may have been discovered recently – you may like to check.

2 Compare the surface temperatures of Jupiter, Saturn, Uranus, Neptune and Pluto.
   a) Can you see a trend?
   Explain your answer.
   b) Which planet temperature appears unexpected?
   Explain why it is unexpected.
   c) Suggest a reason for the unexpected data.

3 What kind of spacecraft would you need to land and survive on
   a) Mercury,
   b) Venus?

Table 27.3 More planet data.

Planet	Distance from Sun in million km (approx.)	Range of surface temperature in °C	Number of moons	Composition of atmosphere
Mercury	58	−180–450	0	–
Venus	108	480	0	carbon dioxide 98% nitrogen 2% (pressure of atmosphere is 90 times greater than on Earth)
Earth	150	−89–58	1	nitrogen 78% oxygen 20% other gases
Mars	228	−120–25	2	carbon dioxide (pressure of atmosphere is 0.7% of that on Earth)
Jupiter	778	−140	at least 39 (16 major moons)	hydrogen 90% helium 10%
Saturn	1427	−180	at least 30 (18 major moons)	hydrogen, helium and some ammonia
Uranus	2871	−210	at least 22	hydrogen, helium and methane
Neptune	4497	−210	at least 8	hydrogen, helium and methane
Pluto	5914	−230	1	–

28 A closer look at heat energy

A great deal of the energy we use in the home is used to provide heat for cooking, for heating water for washing ourselves and our clothes, and for keeping us warm. Heat energy (thermal energy) is also produced in significant quantities whenever energy is converted from one form to another.

Heat and temperature

The hotness or coldness of a substance is measured by taking its temperature. The temperature of a substance is measured on a scale which has two fixed points. The most widely used temperature scale is the Celsius scale. Its two fixed points are 0 °C (the melting point of ice or freezing point of water) and 100 °C (the boiling point of water). In between the two fixed points the scale is divided into one hundred units or degrees. The scale may be extended below 0 °C and above 100 °C; laboratory thermometers usually have a scale reading from −10 °C to 110 °C.

The thermometer compares the temperature of the substance in which the bulb is immersed with the freezing point and boiling point of water. It compares the hotness or coldness of a substance. It does not measure the total internal energy (see page 232) of the substance.

The lowest possible temperature, known as absolute zero, is −273 °C. Temperatures can go as high as millions of degrees Celsius.

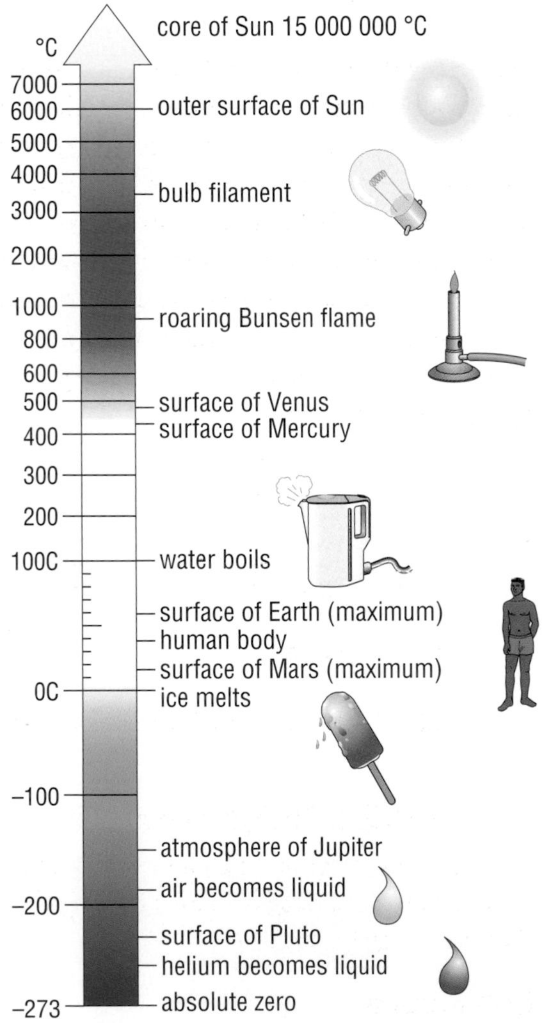

1 How much hotter is
  a) 45 °C than 30 °C,
  b) 20 °C than −15 °C?
2 Why are two fixed points needed for a temperature scale and not just one?

°C

core of Sun 15 000 000 °C

7000 —
6000 — — outer surface of Sun
5000 —
4000 —
3000 — — bulb filament
2000 —
1000 — — roaring Bunsen flame
800 —
600 —
500 — — surface of Venus
400 — — surface of Mercury
300 —
200 —
100C — — water boils
         — surface of Earth (maximum)
         — human body
         — surface of Mars (maximum)
0C — — ice melts
−100 —
         — atmosphere of Jupiter
         — air becomes liquid
−200 —
         — surface of Pluto
         — helium becomes liquid
−273 — — absolute zero

Figure 28.1 The Celsius scale of temperature.

Heat and internal energy

The 'heat' in a substance is really a measure of the total kinetic energy of the atoms and molecules of a substance, due to its internal energy. The total amount of heat in a substance is related to its mass. A large mass of a substance holds a larger amount of heat – it has more internal energy – than a smaller mass. For example, if 100 cm³ of water is heated in a beaker with a Bunsen burner on a roaring flame it will take less time to reach 100 °C than 200 cm³ of water would because it has a smaller mass.

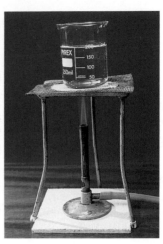

Figure 28.2 When heating two masses of water, more heat energy needs to be supplied to the larger mass to reach the same temperature.

When a substance is heated the (thermal) energy supplied increases the internal kinetic energy which means the atoms and molecules in the substance move faster and further. If the temperature of the substance is taken with a thermometer, kinetic energy from the substance passes to the atoms or molecules from which the thermometer liquid is made and causes them to move faster too. This leads to an expansion of the liquid in the thermometer tube. The temperature reading is a measure of the (average) kinetic energy of the particles hitting the bulb, but not of the total kinetic energy of all the particles in the substance.

3 Why does it take a full kettle longer to boil than a half-full kettle?

4 Which do you think contains more internal energy, a teaspoon of boiling water or a pan full of water at 50 °C?

a)   **b)**

Figure 28.3 Particles in motion in **a)** a cool substance and **b)** a hot substance.

How heat energy travels

There are three ways in which heat energy can travel. They are conduction, convection and radiation. Together they are known as thermal energy transfer.

Conduction

The heat energy is passed from one particle of a material to the next particle. For example, when a metal pan of water is put on a hot plate of a cooker the atoms in the metal close to the hot plate receive heat energy and vibrate more vigorously. They knock against the atoms a little further into the bottom of the pan and make them vibrate more strongly too. These atoms knock against other atoms a little further up

Figure 28.4 The conduction of heat through the bottom of a pan.

5 A metal rod had drawing pins stuck to it with wax and was heated at one end as shown in Figure 28.5.

Figure 28.5

a) What do you think will happen in this experiment? Explain your answer.

b) How could this experiment be adapted to compare the conducting properties of different materials?

6 Imagine that a football represents heat energy and football players represent particles in a material. Which of the following events is like

a) conduction,

b) convection?

Explain your answers.

i) The players pass the ball to each other to move it up the field.

ii) A defender receives the ball and runs up field with it into an attacking position.

7 When coal burns, particles of soot rise up above the fire and make smoke. Why doesn't the smoke move along the ground?

and the kinetic energy is passed on. Eventually the inner surface of the pan, which is next to the water, becomes hot too.

Conduction can occur easily in solids, less easily in liquids but hardly at all in gases because the gas atoms are too far apart to affect each other. It cannot occur in a vacuum, such as outer space, where there are no particles to pass on the heat energy. Conduction is fastest in metals because they have electrons that are free to move. When a metal is heated the electrons in that part move about faster and pass on heat energy to nearby electrons and atoms, so that the heat energy spreads quickly through to cooler parts of the metal.

Materials that allow heat to pass through them easily are called conductors of heat. Materials that do not allow heat to pass through them easily are called insulators.

Insulators are useful in reducing the loss of heat energy. For example, the fabric of a thick woollen pullover is a good insulator. It keeps in your body heat in cold conditions. Insulation materials are also used on the floor of lofts to reduce the escape of heat through the roof of a home.

Convection

The heat energy is carried away by the particles of the material changing position. For example, the water next to the hot surface at the bottom of the pan receives heat from the metal. The molecules of water next to the metal move faster and further apart as their kinetic energy increases. This makes the water next to the pan bottom less dense than the water above it and so this warm water rises. Cooler water from above moves in to take the place of the rising warmer water. The cool water is also warmed and rises. It is replaced by yet more cool water and convection currents are set up as shown in Figure 28.6.

Figure 28.6 The convection currents in a pan of water heated from below.

Convection can only occur in liquids and gases, not in solids where the particles are not free to move about, nor in a vacuum such as outer space.

Radiation

Energy can travel through air or through a vacuum as electromagnetic waves (see page 233). For example, as the pan of water gets hotter you can put your hand near its side and feel the heat on your skin even though you are not touching the metal. The sides of the pan are radiating infrared waves. These carry the heat energy from the surface of the pan to the surface of your skin, which is warmed by them.

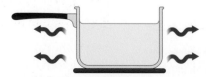

Figure 28.7 Heat radiation from a hot pan.

8 a) The temperature of the land surface is higher than that of the sea surface during the day. Use the ideas of convection currents to suggest what happens to air above the land and above the sea. Which way do you think the wind will blow across the promenade in Figure 28.8?

Figure 28.8

b) At night the land surface is cooler than the sea surface. Does this affect the wind direction? Explain your answer.

9 In question 6 the energy was transferred by particles (players). How is the transfer of heat energy by radiation different?

10 How does the type of surface of an object affect the way it radiates and absorbs heat energy?

All objects radiate infrared, but the hotter the object the more infrared energy it radiates and the shorter the wavelength of the waves.

Some infrared radiation can pass through certain solids such as glass. For example, the infrared radiation from the Sun can pass through glass in a greenhouse but the (longer wavelength) infrared radiation from the ground and the plants inside the greenhouse cannot pass back out through the glass. This infrared radiation is trapped and warms the contents of the greenhouse.

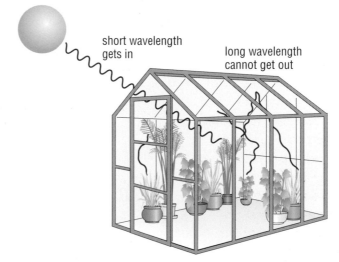

short wavelength gets in

long wavelength cannot get out

Figure 28.9   A greenhouse traps infrared energy.

The type of surface affects the amount of heat energy radiated from an object in a given time. Darker colours radiate energy more rapidly than lighter colours, and black surfaces radiate the most rapidly. The surfaces which radiate the energy least rapidly are light shiny surfaces, like the surface of polished metal.

The type of surface also affects the amount of radiated energy absorbed by the surface in a given time. For example, a light shiny surface absorbs energy the least rapidly, while a black surface absorbs energy the most rapidly.

The Thermos flask

In 1892 Sir James Dewar invented a special insulating flask called a Dewar flask. The Dewar flask is now more widely known as a vacuum or Thermos flask and is used mainly to keep drinks hot. Its construction is shown in Figure 28.10 (overleaf).

The walls of the flask are made of glass, which is a poor conductor of heat, and are separated by a vacuum. The glass walls themselves have shiny surfaces. The surface of the inner wall radiates very little heat and the surface of the outer wall absorbs very little of the heat that is radiated from the inner wall. The cork supports are poor conductors of heat and the stopper prevents heat being lost by convection and evaporation in the air above the surface of the liquid.

11 Which forms of energy transfer does the vacuum prevent?

12 Why are the glass walls shiny? How would the efficiency of the flask be affected if the walls were painted black? Explain your answer.

13 How could a warm liquid lose heat if the stopper was removed? Explain your answer.

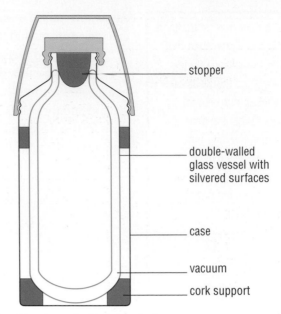

stopper

double-walled glass vessel with silvered surfaces

case

vacuum

cork support

Figure 28.10   Structure of a Thermos flask.

Summary

- Temperature is a measure of the hotness or coldness of a substance (*see page 282*).
- When a substance is heated its internal energy increases and its particles move faster (*see page 282*).
- A thermometer measures the average kinetic energy of the particles hitting the thermometer bulb (*see page 282*).
- Heat energy is passed from particle to particle by conduction (*see pages 283–284*).
- Materials can be divided into conductors and insulators according to how easily heat passes through them (*see page 284*).
- Heat energy is carried by moving particles in a convection current (*see page 284*).
- Heat energy is carried by electromagnetic waves in radiation (*see page 284*).

End of chapter questions

An investigation was carried out to see if useful amounts of heat energy from the Sun could be trapped in trays of water. Three metal trays were used. Each one was 25 cm long, 20 cm wide and 5 cm deep and was filled with 1500 cm³ of water.

Tray 1 had a glass plate cover and the water in it was untreated.
Tray 2 had some black ink added to the water before the glass plate cover was put over it.
Tray 3 had some black ink added to the water before the glass plate cover was put over it, then the sides and base were packed with vermiculite – a spongelike, rocky material.

The trays were exposed to sunlight during the day for seven hours and the air temperature and the temperature of the water in each tray were taken every hour. Table 28.1 shows the data that were collected.   *(continued)*

Table 28.1

Time	Air temperature (°C)	Water temperatures (°C)		
		Colourless (without insulation)	Black (without insulation)	Black (with insulation)
11.15 am	18.7	17.5	17.5	17.5
12.15 pm	19.0	18.9	20.0	20.4
1.15 pm	20.5	22.4	25.1	26.0
2.15 pm	18.3	21.0	23.0	24.9
3.15 pm	18.9	21.5	23.4	25.6
4.15 pm	20.6	24.5	28.3	29.1
5.15 pm	20.2	28.5	31.8	33.2
6.15 pm	19.2	26.0	28.1	31.5

1 Did the water fill the trays to the top? Explain your answer.

2 Plot lines of the data for the air temperature and the temperature of each tray, on the same graph.

3 Compare the graphs you have drawn.

4 What was the purpose of the ink and the vermiculite? Explain your answers.

5 What is the maximum temperature rise? When was it achieved and in which tray?

6 What can you conclude from this investigation?

29 Magnetism

Three metallic elements show strong magnetic properties. They are iron, cobalt and nickel. Steel is a metal alloy which can show magnetic properties. It is made from iron and carbon. Steel can also be mixed with other metals to make an alloy which does not show magnetic properties. For example, stainless steel is made from steel, chromium and nickel and it does not show magnetic properties.

Materials that show magnetic properties do not show them all the time. For example, steel paper clips do not generally attract and repel each other. When a material is showing magnetic properties it is said to be magnetised and is known as a magnet. The most widely used magnets used to be made from steel but most magnets are now made of mixtures of the magnetic metals. Alnico is an example.

It is thought that the word 'magnet' comes from the name of the ancient country of Magnesia which is now part of Turkey. In this region large numbers of black stones were found which had the power to draw pieces of iron to them. The black stone became known as lodestone or leading stone because of the way it could be used to find directions. Today it is known as the mineral magnetite and it has been found in many countries.

1 Which three metals do you think might be present in Alnico? Explain your answer. Which ones are magnetic?

Figure 29.1 Magnetite is a naturally occurring magnet.

2 How do magnetic materials differ from non-magnetic materials in
   a) what they are made of,
   b) their properties?

The behaviour of magnets

Magnets can attract or repel other magnets and can attract any magnetic material even if it is not magnetised. When suspended from a thread, a bar magnet aligns itself in a north–south direction.

Non-magnetic materials, such as wood, paper, plastic and most metals, cannot be magnetised and so can do none of these things. Some, such as paper and water, can let the force of magnetism pass through them while other materials, such as a steel sheet, do not let the force of magnetism pass through them.

The strength of the magnetic force

At each end of a bar magnet is a place where the magnetic force is stronger than at other places in the magnet. These places where the magnetic force is strongest are called the poles of the magnet. The end of the magnet which points towards north when the magnet is free to move is called the north-seeking pole or north pole. At the other end of the magnet is the south-seeking pole or south pole.

When the north pole of one magnet is brought close to the south pole of another magnet that is free to move, the south pole moves towards the north pole. Similarly, a north pole is attracted to a south pole. However, two south poles repel each other, as do two north poles. These observations can be summarised by the phrase 'different poles attract, similar poles repel'.

If you bring a steel paper clip (which is not magnetised) towards either pole of a magnet you will feel the pull of the magnetic force become stronger as the paper clip gets closer to the pole. As you move the paper clip away again you will feel the pull of the magnet become weaker.

3 What is the relationship between the distance from the magnet and the strength of the magnetic force?

The magnetic field

The region around a magnet in which the pull of the magnetic force acts on magnetic materials is called the magnetic field.

The field around a magnet can be shown by using a piece of card and iron filings. The card is laid over the magnet and the iron filings are sprinkled over the paper. Each iron filing has such a small mass that it can be moved by the magnetic force of the magnet if the paper is gently tapped. The iron filings line up as shown in Figure 29.2. The pattern made by the iron filings is called the magnetic field pattern.

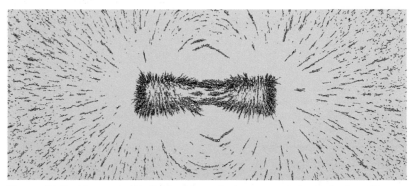

Figure 29.2   The magnetic field pattern of a bar magnet shown by iron filings.

The iron filings appear to form lines around the magnet. This phenomenon can be checked by using a plotting compass and a piece of paper and pencil. The magnet is placed in the centre of the paper and the plotting compass is placed on one side of the magnet close to its north pole. The north pole of the compass will point away from it. The position of the north pole of the compass is marked on the paper and the plotting compass is then moved so that its south pole is over the mark made on the paper. The position of the north pole is marked again with the plotting compass in the new position and the process is repeated until the plotting compass reaches the south pole of the magnet. If the points marking the positions of the north pole of the compass are joined together by a line running from the north pole to the south pole of the magnet (Figure 29.3a), this will represent one of the magnetic 'lines of force' forming the field pattern. Arrows should be drawn on the lines, pointing from the magnet's north pole to its south pole (Figure 29.3b).

4 How does the information from the activity with the plotting compass compare with the field pattern produced by the iron filings?

5 Figure 29.4 shows iron filings spread out when in contact with the end of a bar magnet. Make a drawing of how you think the field lines are arranged all around the magnet.

Figure 29.4

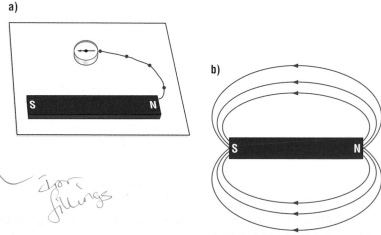

Figure 29.3   **a)** Drawing a magnetic line of force and **b)** the magnetic field pattern around a bar magnet.

The Earth's magnetic field

At the centre of the Earth is the Earth's core. It is made from iron and nickel and is divided into two parts – the inner core made of solid metal and the outer core made of liquid metal. As the Earth spins the two parts of the core move at different speeds and this is thought to generate the magnetic field around the Earth and make the Earth seem to have a large bar magnet inside it.

The Earth spins on its axis which is an imaginary line that runs through the centre of the planet. The ends of the line are called the geographic north and south poles. Their positions on the surface of the Earth are fixed. Magnetic north – towards which the free north pole of a magnet points – is not at the same place as the geographic north pole (Figure 29.5), and it changes position slightly every year.

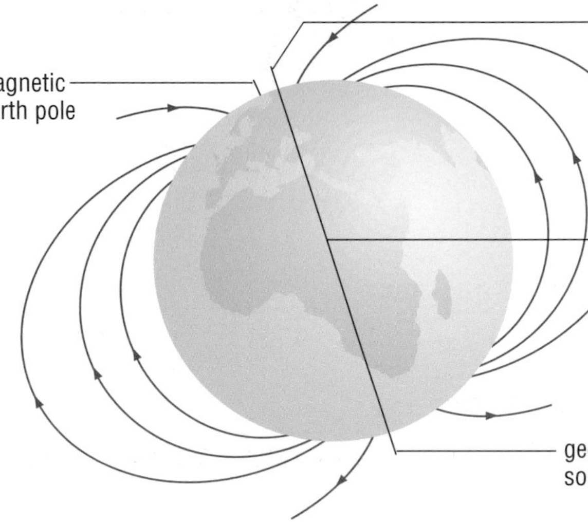

6 a) Look at the field pattern around the Earth in Figure 29.5. Which pole of the imaginary bar magnet inside the Earth coincides with magnetic north?

b) Draw a bar magnet inside the Earth and label its poles. Also label the position of the south magnetic pole on the Earth's surface.

Figure 29.5 The Earth's geographic and magnetic poles do not coincide.

The north magnetic pole originally got its name because it is the place to which the north poles of bar magnets point. In reality it is the Earth's south magnetic pole because it attracts the north poles of magnets.

Similarly the south magnetic pole is really the Earth's north magnetic pole because it attracts the south poles of bar magnets. However for most purposes the old, and incorrect, names for the magnetic poles are still used.

The link between magnetism and electricity

Hans Christian Oersted (1777–1851) was a Danish physicist who studied electricity. In one of his experiments he was passing an electric current along a wire from a voltaic pile (an early type of battery) when he noticed the movement of a compass needle which had been left near the wire. This chance observation led to many discoveries of how magnetism and electricity are linked together, and many modern applications.

When an electric current passes through a wire it generates a magnetic field around the wire. A compass can be placed at different places on a card around the wire, as shown in Figure 29.6, and lines of force can be plotted.

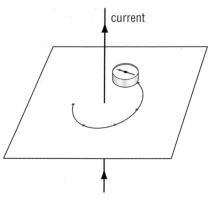

Figure 29.6 Plotting magnetic field lines around a current-carrying wire.

When the current flows up through the card, the field shown in Figure 29.7a is produced. When the current flows down through the card, the field shown in Figure 29.7b is produced.

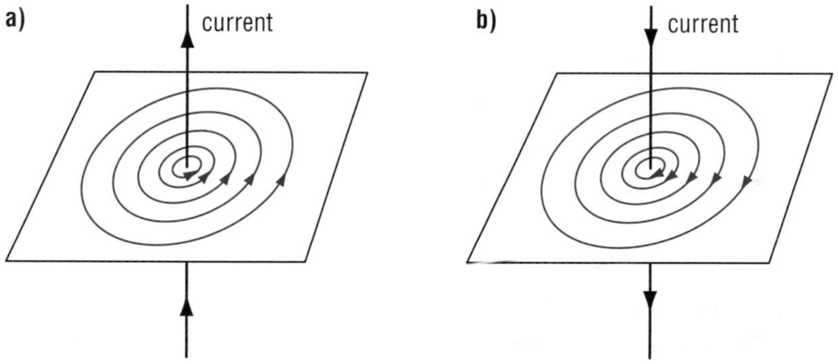

a) current   b) current

Figure 29.7 The magnetic field around a current-carrying wire.

7 How are the fields in Figures 29.7a and b different?
8 How does the strength of the magnetic field around the wire vary?

Lines of force on diagrams of magnetic fields show not only the direction of the field as given by a plotting compass but also the strength of the field in different places. The lines of force are close together where the field is strong and wider apart where the field is weaker.

If the wire is made into a coil and connected into a circuit, a magnetic field is produced around the coil as shown in Figure 29.8.

If a piece of steel is placed inside the wire coil and the current is switched on, the magnetism of the coil and the steel is stronger than that of the coil alone. The current flowing through the coil induces magnetism in the steel. Steel is magnetically 'hard'. This means that when the current is switched off the steel keeps some of the magnetism it acquired.

9 Compare the magnetic field of a bar magnet (Figure 29.3b, page 289) with that produced by a current in a wire coil (Figure 29.8).

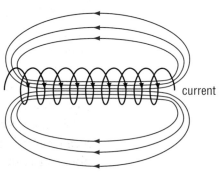

current

Figure 29.8 The magnetic field around a current-carrying coil.

If a piece of iron is placed inside the coil it makes an even stronger magnet when the current is switched on than the steel did. However, iron is magnetically 'soft'. This means that when the current is switched off the iron loses its magnetism completely.

The electromagnet

An electromagnet is made from a coil of wire surrounding a piece of iron. When a current flows through the coil, magnetism is induced in the iron, and the coil and iron form a strong electromagnet. When the current is switched off the electromagnet loses its magnetism completely, straight away. This device, which can instantly become a magnet then instantly lose its magnetism, has many uses. For example, a large electromagnet is used in a scrapyard to move the steel bodies of cars (Figure 29.9).

10 Describe how you think an electromagnet can be used to make a stack of scrapped cars three cars high.

Figure 29.9 An electromagnet in use in a scrapyard.

The reed switch

A reed switch (Figure 29.10) is a magnetic switch. It has two pieces of soft iron, called the reeds, supported by metal which has a springy property. The reeds are enclosed in a glass container which is filled with an inert gas. This gas is used instead of air because the metal does not react with it and so does not corrode.

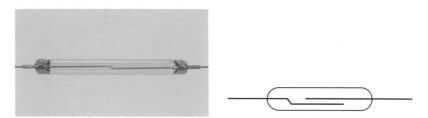

Figure 29.10 A reed switch and its symbol.

When a magnet is brought close to the reed switch it makes the soft iron reeds become magnets (Figure 29.11). The opposite poles on the free ends of the reeds attract each other. The magnetic force between them bends the two pieces of springy metal so that the two reeds touch and close the circuit, allowing the current to flow. When the magnet is taken away the soft iron reeds lose their magnetism and the tension force in the springy metal pulls the reeds apart.

Figure 29.11   How a reed switch works.

This type of reed switch is used in burglar alarms, where the reed switch is placed in the door frame and magnets are placed in the door and the door frame (Figure 29.12).

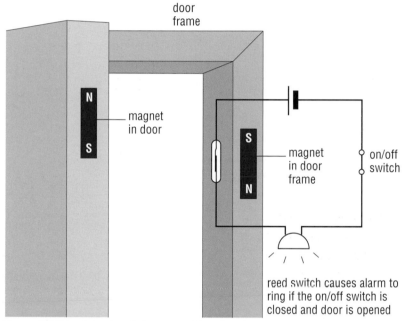

door frame

magnet in door

magnet in door frame

on/off switch

reed switch causes alarm to ring if the on/off switch is closed and door is opened

Figure 29.12   A reed-switch operated burglar alarm.

The reed relay

A reed relay is a reed switch like the one shown in Figure 29.10 but with a coil of wire wrapped round it.

Figure 29.13   A reed relay and its symbol.

When an electric current flows through the coil, a magnetic field is produced in and around the coil which causes the reeds to move. The coil and reeds are connected in separate circuits. Only a small current is needed by the coil to move the reeds but a large current can be conducted in the other circuit by the reeds. This allows a circuit through which a large current passes to be controlled by a circuit in which a small current flows.

A motor can be controlled by light by using the circuits shown in Figure 29.14. When it is dark, the high resistance of the LDR allows only a very small current to flow in circuit A. It is not enough to produce a magnetic field around the reeds in the relay. When it is light the resistance of the LDR falls and the size of the current increases. This produces a magnetic field around the reeds which makes them come together and close circuit B. A much larger current flows through circuit B to make the motor work.

The use of the reed relay allows a component which only requires a small current to control a component which requires a large current.

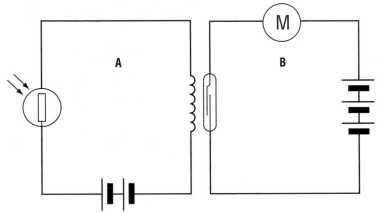

Figure 29.14 A light-operated motor circuit using a reed relay.

13 What happens in circuits A and B when it becomes darker again?

The electromagnetic relay

An electromagnetic relay is a switch operated by an electromagnet. When the driver of a car puts the key in the dashboard and twists it, the ignition switch is turned on using an electromagnetic relay. A small current from the car's battery passes through the coil of an electromagnet and the magnetic force of the electromagnet pulls on a piece of iron called an armature (Figure 29.15, opposite). An armature is a part of an electrical device which moves when a magnetic field develops around it. The magnetic force is strong enough to overcome the forces in the attached spring and it is squashed. The armature moves towards the electromagnet and the end of the armature pushes down on the contacts of springy metal. The contacts touch and close the switch in a circuit through which a very large current flows. This current is used to turn the starter motor which starts the engine running. The big current only goes a short distance. It does not need to go all the way to the key and back.

14 When the driver hears the engine start running he or she stops twisting the ignition key and the current no longer flows to the relay's electromagnet. What happens to the
a) spring,
b) armature,
c) contacts?

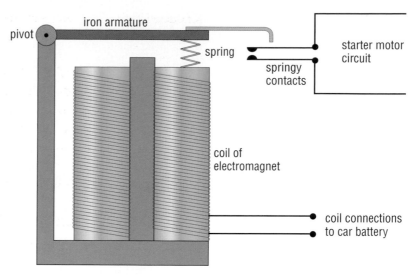

Figure 29.15 An electromagnetic relay operating a starter motor circuit.

The electric bell

Look at Figure 29.16 and see if you can work out the path the current takes through the circuit when the switch is pushed.

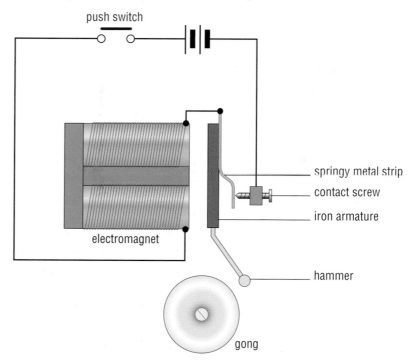

Figure 29.16 The circuit of an electric bell.

When the switch is pushed the current passes through the coil and the electromagnet pulls the armature to it. This makes the hammer strike the gong. When the armature is pulled to the electromagnet a gap develops between the springy metal strip and the contact screw, and the circuit is broken. The current stops flowing and the electromagnet loses its magnetism. This makes the armature swing back to its original position. The springy metal strip and the contact screw now touch again and complete the circuit so the armature is pulled to the electromagnet once more. The bell is made to ring by the repeated beating of the hammer until the push switch is released.

15 Describe the changes that take place in the springy metal strip holding the armature when the current
a) flows,
b) stops flowing.
Use the term strain force in your description.

MAGNETISM

295

Summary

- Magnetic materials are attracted by a magnet; non-magnetic materials are not (*see page 288*).
- A magnet can attract or repel another magnet (*see page 288*).
- A bar magnet aligns itself in a north–south direction when it is free to move (*see page 288*).
- A magnet has a north-seeking pole and a south-seeking pole (*see page 288*).
- A magnetic field exists around a magnet (*see page 289*).
- The Earth has a magnetic field (*see page 290*).
- A wire with an electric current passing through it has a magnetic field around it (*see page 291*).
- An electromagnet is a magnet whose magnetism can be switched on and off by switching a current on and off (*see page 292*).
- A reed switch is opened and closed by a magnet (*see page 292*).
- A reed relay is a reed switch that is controlled by a current-carrying coil (*see page 293*).
- An electromagnetic relay is a switch operated by an electromagnet (*see page 294*).
- An electromagnet is used to produce the repeated ringing of an electric bell (*see page 295*).

End of chapter questions

1 Why do magnets line up in a north–south direction?

2 Assess the importance of magnetism in the working of electrical devices in everyday life.

30 *Light*

Light is a form of energy. It is a form of electromagnetic radiation. Objects that emit light are said to be luminous while those that do not emit light are said to be non-luminous. Non-luminous objects can only be seen if they are reflecting light from a luminous source. The Moon is a non-luminous body – the 'moonlight' it produces is reflected sunlight.

Most luminous objects, such as the Sun, stars, fire and candle flames, release light together with a large amount of heat, that is, infrared radiation.

Figure 30.1   A bonfire is luminous: it radiates light and heat.

1 What is the luminous object which is providing light for you to read this book?

Light rays

Light leaves the surface of a luminous object in all directions but if some of the light is made to pass through a hole it can be seen to travel in straight lines. For example, when sunlight shines through a small gap in the clouds it forms broad sunbeams with straight edges (Figure 30.2, overleaf). The path of the light can be seen because some of it is reflected from dust in the atmosphere. Similarly, sunlight shining through a gap in the curtains of a dark room produces a beam of light which can be seen when the light reflects from the dust in the air of the room.

Figure 30.2 Although the Sun radiates light in all directions, the sides of sunbeams seem almost parallel because the Sun is a very distant luminous object.

Smaller lines of light, called rays, can be made by shining a lamp through slits in a piece of card.

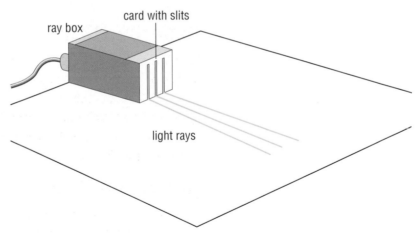

Figure 30.3 Making rays of light.

Classifying non-luminous objects

Non-luminous materials can be classified as transparent, translucent or opaque according to the way light behaves when it meets them. When light shines on a transparent material, such as glass in a window, it passes through it and so objects on the other side of it can be seen clearly.

When light shines on a translucent object, such as tracing paper, some of the light passes through but many light rays are scattered. Objects on the other side cannot be seen clearly unless they are very close to the translucent object.

When light shines on an opaque object none of the light passes through it.

Shadows

When a beam of light shines on an opaque object the light rays which reach the object are stopped while those rays which pass by the edges continue on their path. A region without light, called a shadow, forms behind the object. The shape of the shadow may not be identical to the shape of the object because the shadow's shape depends on the position of the light source and on where the shadow falls.

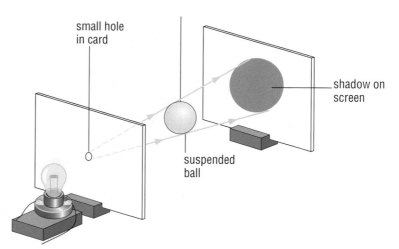

Figure 30.4 Formation of a shadow.

For discussion

How might the shadow of a brick appear if light travelled in a curve from the light source?

Shadows are also cast by the Moon and the Earth (see pages 275–276).

Reflecting light

Your bedroom is probably full of objects but if you were to wake in the middle of the night you could not see them clearly because they are not luminous. You can only see them by reflected light and unless your room is partially lit by street lights the objects will not be clearly seen until sunrise. The way light is reflected from a surface depends on whether the surface is smooth or rough.

Studying reflections

A few terms are used in the study of light which make it easier for scientists to describe their investigations and ideas. In the study of reflections the following terms are used:

● incident ray – a light ray that strikes a surface
● reflected ray – a light ray that is reflected from a surface
● normal – a line perpendicular (that is at 90°) to the surface where the incident ray strikes
● angle of incidence – the angle between the incident ray and the normal
● angle of reflection – the angle between the reflected ray and the normal
● plane mirror – a mirror with a flat surface
● image – the appearance of an object in a smooth, shiny surface. It is produced by light from the object being reflected by the surface.

The way the incident ray, the normal and the reflected ray are represented diagrammatically is shown in Figure 30.5. The back surface of a mirror is usually shown as here, as a line with short lines at an angle to it.

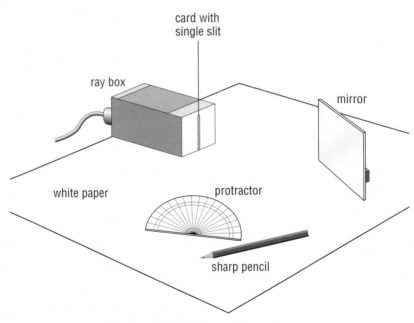

Figure 30.5   Reflection of light from a plane mirror.

The way light rays are reflected from a plane mirror can be investigated using the equipment shown in Figure 30.6.

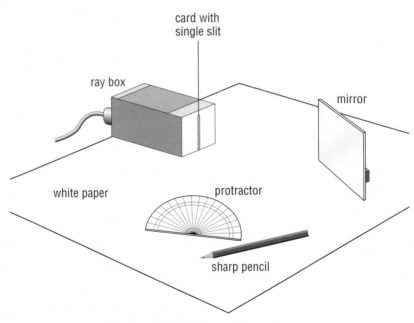

Figure 30.6   Investigating reflection from a plane mirror.

Objects with smooth surfaces

Glass, still water and polished metal have very smooth surfaces. Light rays striking their flat surfaces are reflected as shown in Figure 30.8, opposite. The angle of reflection is equal to the angle of incidence. When the reflected light reaches your eyes you see an image (Figure 30.9).

2 Figure 30.7 shows three drawings made of the path of incident and reflected rays in an experiment using the apparatus in Figure 30.6. Use a protractor to measure the angle of incidence and angle of reflection. What do these drawings tell you about the process of reflection?

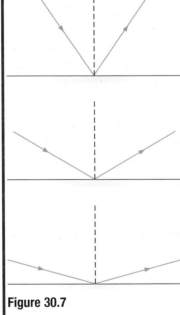

Figure 30.7

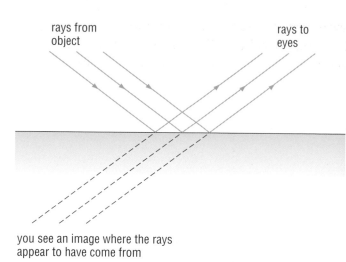

rays from object

rays to eyes

you see an image where the rays appear to have come from

Figure 30.8 Regular reflection from a smooth surface.

Figure 30.9 Light reflected from the smooth surface of a lake can produce an image in the water.

Figure 30.10 Your image in a mirror is the wrong way round.

Two kinds of images

There are two kinds of images that can be formed with light. They are real images, such as those produced on a cinema screen by projector lenses, and virtual images, which cannot be projected onto a surface but only appear to exist, such as those in a plane mirror or other smooth, shiny surface.

The virtual image of yourself that you see when you look in a plane mirror is the same way up as you are, is the same size as you are, and is at the same distance from the mirror's surface as you are but behind the mirror instead of in front of it. The main difference between you and your virtual image is that the virtual image is the 'wrong way round' – for example, your left shoulder appears to be the right shoulder of your virtual image.

The periscope

Two plane mirrors may be used together to give a person at the back of a crowd a view of an event.

Figure 30.11 Some of the people in this scene are using periscopes to help them see over the crowd.

3 Copy Figure 30.12 and draw in the path of a ray of light travelling from the golfer to the eye.

4 Why is a periscope useful on a submarine?

The arrangement of the mirrors in a periscope is shown in Figure 30.12.

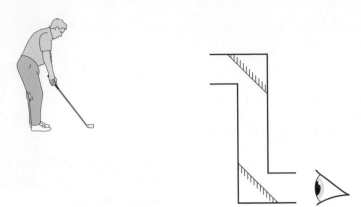

Figure 30.12 A simple periscope.

Objects with rough surfaces

Most objects have rough surfaces. They may be very rough like the surface of a woollen pullover or they may be only slightly rough like the surface of paper. When light rays strike any of these surfaces the rays are scattered in different directions as Figure 30.13 shows.

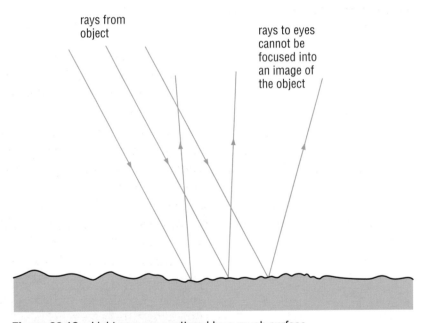

rays from object

rays to eyes cannot be focused into an image of the object

Figure 30.13 Light rays are scattered by a rough surface.

You see a pullover or this page by the light scattered from its surface. You do not see your face in a piece of paper because the reflection of light is irregular, so it cannot form an image.

Passing light through transparent materials

If a ray of light is shone on the side of a glass block as shown in Figure 30.14a, the ray passes straight through, but if the block is tilted the ray of light follows the path shown in Figure 30.14b.

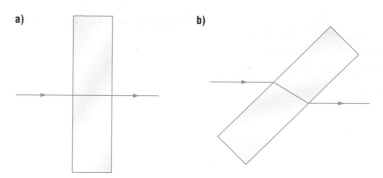

Figure 30.14 Light is refracted if the incident ray is not at 90° to the surface of the transparent material.

This 'bending' of the light ray is called refraction. The angle that the refracted ray (see Figure 30.15) makes with the normal is called the angle of refraction.

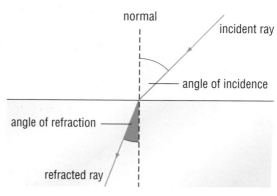

Figure 30.15 The angle of incidence and the angle of refraction.

The refraction of light as it passes from one transparent substance or 'medium' to another is due to the change in the speed of the light. Light travels at different speeds in different media. For example, it travels at almost 300 million metres per second in air but only 200 million metres per second in glass. If the light slows down when it moves from one medium to the other, the ray bends towards the normal. If the light speeds up as it passes from one medium to the next, the ray bends away from the normal.

Light speeds up as it leaves a water surface and enters the air. A light ray appears to have come from a different direction than that of the path it actually travelled (see Figure 30.16a and b). The refraction of the light rays makes the bottom of a swimming pool seem closer to the water surface than it really is. It also makes streams and rivers seem shallower than they really are and this fact must be considered by anyone thinking of wading across a seemingly shallow stretch of water. The refracted light from a straw in a glass of water makes the straw appear to be bent.

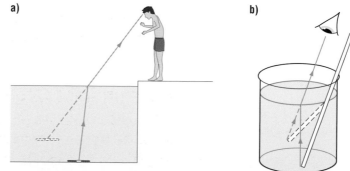

Figure 30.16 Refraction of light as it passes from water to air makes an object appear closer to the surface than it really is.

5 How is the reflection of a light ray from a plane mirror (see page 300) different from the refraction of a light ray as it enters a piece of glass?

The prism

A triangular prism is a glass or plastic block with a triangular cross-section.

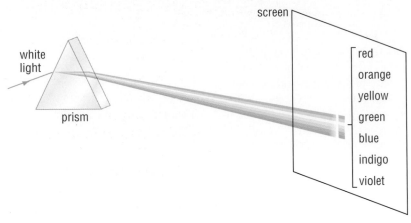

Figure 30.17 White light passing through a prism is split up into its constituent colours, forming a spectrum.

When a ray of sunlight is shone through a prism at certain angles of incidence and its path is stopped by a white screen, a range of colours, called a spectrum, can be seen on the screen.

Light behaves as if it travelled as waves. The 'white' light from the Sun contains light of different wavelengths which give different coloured light. When they pass through a prism the light waves of different wavelengths travel at slightly different speeds and are spread out, by a process called dispersion, to form the colours of the spectrum. The light waves with the shortest waves are slowed down or refracted the most.

The rainbow

If you stand with your back to the Sun when it is raining or you look into a spray of water from a fountain or a hose you may see a rainbow. It is produced by the refraction and reflection of the Sun's light through the water drops. Figure 30.18 shows the path of a light ray and how the colours in it spread out to form the order of colours – the spectrum – seen in a rainbow.

Sometimes a second, weaker rainbow is seen above the first because two reflections occur in each droplet. In the second rainbow the order of colours is reversed.

> **6** Look at Figure 30.17. Which colour of light has the shortest wavelength? Explain your answer.

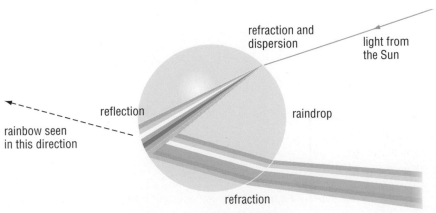

Figure 30.18 Formation of a rainbow.

Summary

- Light is a form of energy that is released from luminous objects (*see page 297*).
- Materials can be classified as opaque, translucent or transparent (*see page 298*).
- A shadow forms when light rays are stopped by an opaque object (*see page 299*).
- We see non-luminous objects by the light they reflect (*see page 299*).
- Light rays are reflected from a smooth surface at the same angle at which they strike it (*see page 300*).
- A real image can be formed on a screen but a virtual image cannot (*see page 301*).
- When light rays strike the surface of a transparent material at an angle to the perpendicular they are refracted (*see pages 302–303*).
- A prism can split up sunlight into different colours of light (*see page 304*).

End of chapter question

1 Describe what happens to light in a beam from the time it reaches the Earth from the Sun and shines upon a leaf, to when it enters your eye.

Sound

You have probably performed some experiments on sound without knowing it. At some time most people have made a ruler vibrate by holding one end over the edge of a desk and 'twanging' it. The end of the ruler moves up and down rapidly and a low whirring sound is heard which becomes higher as you pull in the ruler from the edge of the desk.

Figure 31.1 Making a ruler vibrate.

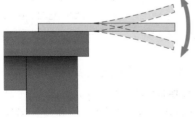

Figure 31.2 Vibration is a to-and-fro movement.

From vibration to sound wave

Any object can make a sound wave when it vibrates. In practical work on sound you might use an elastic band, a guitar string or a tuning fork because they all vibrate easily. A vibration is a movement about a fixed point. This movement may be described as a to-and-fro movement or a backwards and forwards movement (Figure 31.2).

Sound waves can travel in a gas, a liquid or a solid because they all contain particles. When an object vibrates it makes the particles next to it in the gas, liquid or solid vibrate too. For example, when an object vibrates in air it pushes on the air particles around it.

As the vibrating object moves towards the air particles it squashes them together. The particles themselves are not compressed but the pressure in the air at that place rises because the particles are closer together (Figure 31.3a).

a)   b)

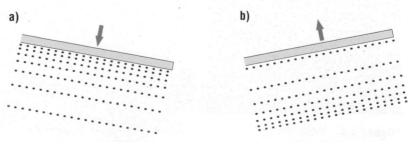

Figure 31.3 A vibrating object causes pressure variations in the air around it.

As the object moves away from the air particles next to it, it gives them more space and they spread out and the pressure at that place falls (Figure 31.3b).

As the object vibrates, the air particles nearby also move backwards and forwards and they in turn cause other air particles further away to squash together and then spread out. This makes alternate regions of high and low pressure which travel through the air away from the vibrating object (Figure 31.4).

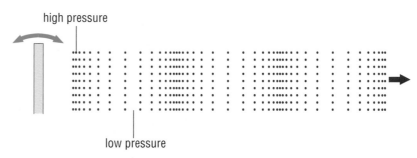

high pressure

low pressure

Figure 31.4 Regions of high and low pressure move away from the vibrating object.

If these changes in pressure were plotted on a graph they would make a waveform similar to that shown in Figure 31.8 (page 308). The waves of sound move out from the vibrating object in all directions.

Sound waves are generated and travel in liquids and solids in the same way as they do in gases. The particles in liquids and solids are held close together by forces of attraction. In a liquid, however, the particles are further apart than in a solid and can move over one another. Sound travels very well through a liquid. It moves faster and further than it does in a gas. The humpback whale emits a series of sounds called songs which travel thousands of kilometres through the ocean. It uses its songs to communicate with other whales.

When sound travels through a solid it moves even faster than through a liquid because of the close interactions of the particles. However, the sound does not travel so far. A snake detects vibrations in the ground with its lower jaw bone. The bone transmits the vibrations to the snake's ears and enables the snake to listen for its prey.

Figure 31.5 These whales communicate by sound waves.

Figure 31.6 This snake is listening for vibrations in the ground.

Sound waves cannot pass through a vacuum because it does not contain any particles. Figure 31.7 shows an experiment that demonstrates this. As air is drawn out of the bell jar with a pump, the sound of the bell becomes quieter. When a vacuum is established in the bell jar the bell cannot be heard although the hammer can be seen striking it.

2 Why is it that a bell in a sealed bell jar
  a) can be heard when the jar is full of air, but
  b) cannot be heard when a vacuum is created in the jar?

Figure 31.7 Sound cannot be heard through a vacuum.

Describing the wave

Figure 31.8 shows the different positions particles can occupy when a sound wave is produced. This type of graph is called a displacement/distance graph.

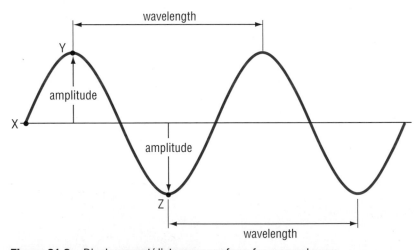

Figure 31.8 Displacement/distance waveform for a sound wave.

A particle at position X in Figure 31.8 is moving through the 'rest' position, a particle at Y has moved the maximum distance in one direction and one at Z has moved the maximum distance in the other direction.

3 Can you think of other ways of describing the wavelength of a wave?

Two characteristics of the wave that can be seen in Figure 31.8 are the amplitude and wavelength. The amplitude is the height of the crest or the depth of the trough and shows the maximum displacement of the particles from their rest position. The wavelength is the distance from the top of one crest to the top of the next crest, or from the bottom of one trough to the bottom of the next trough.

Detecting sound waves

The ear is the organ of the body that detects sound waves. It is divided into three parts – the outer ear, middle ear and inner ear (Figure 31.9).

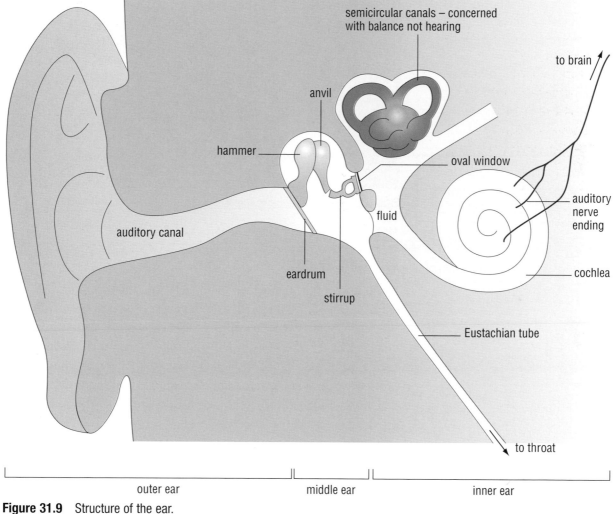

Figure 31.9 Structure of the ear.

The outer ear

When sound waves reach the outer ear some pass directly down the middle of the tube called the auditory canal. Some waves which strike the outer part of the ear are reflected into the auditory canal. At the end of the auditory canal is a thin membrane which stretches across it. This is called the eardrum. When sound waves reach the eardrum they push and pull on it and make it vibrate.

Some predatory animals, such as cats, can turn their outer ears forwards to detect sounds from prey in front of them. Some prey animals, such as rabbits, can turn their outer ears in many directions about their head to listen for approaching predators.

Figure 31.10 This rabbit is able to turn its outer ears to capture sound waves from all directions.

4 Why do people put a hand to their ear when they are listening to someone who is whispering?

The middle ear

In the cavity of the middle ear are three bones. They are called the hammer, anvil and stirrup, after their shapes. The ear bones form a system of levers. When the eardrum vibrates its movements are amplified by the lever system. The oval window on which the stirrup bone vibrates has a much smaller area than that of the eardrum. This difference in area between the eardrum and the oval window causes the vibrations of the eardrum to be amplified as they enter the inner ear and set up vibrations in the fluid there.

The middle ear also has a tube, the Eustachian tube, which connects to the throat. When we swallow, the tube opens and the air in the middle ear is connected to air outside the body. This brief connection allows the air pressure in the ear to adjust to the air pressure outside the body. This balancing of the air pressure allows the eardrum to vibrate as freely as possible.

5 Why do people go partially deaf when they have a very heavy cold and the Eustachian tubes become blocked?

The inner ear

The inner ear is filled with a fluid. The vibrations of the stirrup set up waves in the fluid. There is a membrane with delicate fibres in the cochlea. Each fibre only vibrates in response to a sound wave with a particular pitch (see page 312). When a fibre vibrates it stimulates a nerve ending and a nerve impulse or message is sent to the brain where we become aware of the sound.

The loudness of sounds

The loudness of a sound is related to the movement of the vibrating object. If an object only moves a short distance to and fro from its rest position, it will produce a sound wave with only a small amplitude and the sound that is heard will be a quiet one.

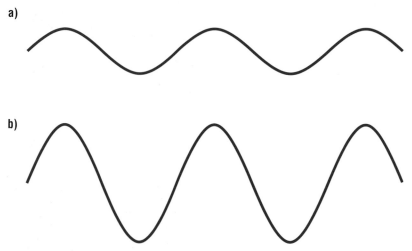

Figure 31.11 Displacement/distance waveforms of **a)** a quiet sound and **b)** a loud sound.

If an object moves a large distance to and fro from its rest position, it will produce sound waves with a large amplitude and the sound that is heard will be a loud one. The loudness of sounds is measured in decibels (see Table 31.1).

Table 31.1 Loudness values of sounds.

Sound	Loudness (decibels)
The sound hurts	140
A jet aircraft taking off	130
A road drill	110
A jet plane overhead	100
A noisy factory floor	90
A vacuum cleaner	80
A busy street	70
A busy department store	60
Normal speech	55
Voices in a town at night	40
A whisper	20
Rustling leaves	10
Limit of normal hearing	0

Figure 31.12 Wearing ear protection in a noisy boiler room prevents ear damage.

6 What kind of ear damage might be caused by a loud explosion? Explain your answer.

For discussion

To prevent ear damage, how should you use earphones on a CD player? Where should you dance at a disco, and where should you sit or stand at a pop concert?

How far do you follow the advice you have given in answer to the above?

Loudness and wave energy

Sound energy passes through the air as the particles move to and fro. When a wave with a small amplitude is generated, a small amount of energy passes through the air. When a wave with a large amplitude is generated a large amount of energy passes. The energy of a sound wave is converted into other forms such as movement energy in the eardrum and ear bones.

Loudness and deafness

The vibrating air particles of a very loud sound can produce such a strong pushing and pulling force on the eardrum that a hole is torn in it. The eardrum is said to be perforated. It no longer vibrates efficiently and the person loses his or her hearing. The eardrum can heal and normal hearing can be restored.

If a person is exposed to a very loud sound or a particular note for a long period of time he or she will no longer be able to hear it. This is due to permanent damage to a nerve ending in the cochlea. People who perform in rock groups are at risk of this kind of deafness, called nerve deafness. In time they may be unable to hear a range of notes which they frequently used in their music. People who work in noisy surroundings, such as airport workers or metal workers in a factory, wear ear protection in the form of ear muffs which cover the ears and reduce the amount of sound energy entering the ears.

A common form of partial deafness, which is not related to the loudness of a sound, is the development of ear wax in the outer ear. This stops sound waves reaching the eardrum. The wax can be removed with warm water under the medical supervision of a nurse.

Some people have growths of tissues in their middle ears which stop the ear bones moving freely. They may be prescribed with a hearing aid. This contains a microphone and amplifier and compensates for some of the loss of amplification that was provided by the ear bones.

The pitch of a sound

You probably have an idea about the pitch of a sound even if you don't know the word. You might describe a sound as a high or a low sound, which really means a high-pitched or a low-pitched sound. For example, when you say 'bing' you are making a higher-pitched sound than when you say 'bong'.

The pitch of the sound an object makes depends on the number of sound waves it produces in a second as it vibrates. This number of waves per second is called the frequency. The frequency of a sound is measured in hertz (abbreviation Hz). The higher the frequency of the wave, the higher the pitch of the sound.

The graphs in Figure 31.13 show the positions that particles occupy at different times as the wave passes. These graphs are called displacement/time graphs. The higher frequency waves have a shorter wavelength than the lower frequency waves. Sound waves share this property with light waves.

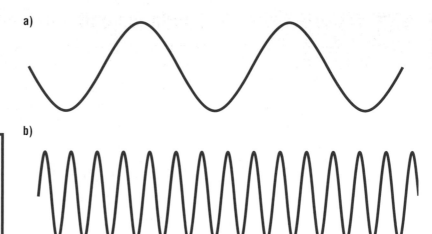

Figure 31.13 Displacement/time waveforms of **a)** a low frequency sound and **b)** a high frequency sound.

7 The following are three frequencies of sound waves: 1800 Hz, 50 Hz, 10 000 Hz
  a) Which has the highest pitch and which has the lowest pitch?
  b) What does Hz stand for?

The ear of a young person is sensitive to frequencies in the range 20 to 20 000 hertz, but the ability to detect the higher frequencies decreases with age. Some people may have a restricted range of hearing due to nerve damage. They may not be able to hear some low-pitched or high-pitched sounds.

The speed of sound

Sound travels at 330 m/s in air at 0 °C, at 343 m/s in air at 20 °C and at 277 m/s in air at 280 °C.

At 20 °C it travels through water at 1500 m/s, through glass at 5000 m/s and through steel at 6000 m/s.

8 How is the speed of sound in air related to the air temperature?

For discussion

Why do you think sound travels at different speeds in solids, liquids and gases?

Why do you think there is the relationship you have described in answering question 8?

Experiments on sound

In the past many scientists have performed experiments to find the speed of sound. Isaac Newton (1642–1727) investigated the speed of sound by measuring the time between a sound being made and its echo from a wall being heard. Other scientists measured the time taken between seeing a distant cannon fire and hearing its sound.

The speed of sound in water was investigated using the apparatus shown in Figure 31.15. The experiment was performed at night. When the lever was pulled down both the arm carrying the bell hammer and the device carrying the match moved.

9 What is an echo?
10 What measurement besides time needs to be taken in all the experiments to determine the speed of sound?

Figure 31.14 Measuring the speed of sound in air.

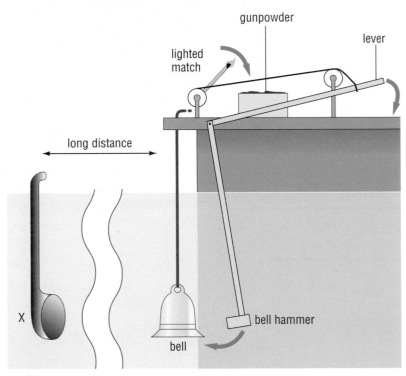

Figure 31.15 Measuring the speed of sound in water.

11 What is the purpose of
  a) the gunpowder, and
  b) the apparatus marked X
    in Figure 31.15?
12 Why do you think the
  experiment to find the speed
  of sound in water was done
  at night?

Summary

- Sounds are made by vibrating objects (*see page 306*).
- Sound travels through materials as waves of vibrating particles (*see page 306*).
- There are three parts to the ear. Each part plays an important role in hearing (*see pages 309–310*).
- The loudness of a sound is related to the amplitude of its waves (*see page 311*).
- The pitch of a sound is related to the frequency of its waves (*see page 312*).
- Sound travels at different speeds in different materials (*see page 313*).

End of chapter question

1 Describe how the vibration of a ruler is detected in the inner part of your ear.

32 *Speed*

Figure 32.1   How do you measure the runners' speed?

A few moments after the finish of this race the times of the runners will be flashed up on a score board. If the race was over 100 metres, or any other distance, you may think that the time gives you the speed that the runner ran that entire distance. However you would be wrong. The runners were stationary when the starter gun fired so they began to run fast or accelerate to get moving. They may have run steadily for most of the race then accelerated as much as they could for the final sprint to the finish. Speed is a steady rate of movement over a distance. To measure the speed of the runners they should have been running hard over the starting line and kept running steadily to the finish.

Calculating speed

The speed at which something moves is the distance travelled in a certain interval of time. The following equation shows how the steady speed of an object can be calculated:

$$\text{speed} = \frac{\text{distance}}{\text{time}}$$

The standard SI unit for speed is m/s. In the laboratory speeds may be measured in m/s or cm/s. The speeds of vehicles may be measured in km/h, although miles per hour (mph) is still commonly used in Britain.

Distance/time graph

The distance travelled by an object over a period of time can be plotted on a graph called a distance/time graph. The distance covered by the object is recorded on the Y axis and the time taken for the object to cover the distance is recorded on the X axis. When a distance/time graph is complete it can be used to study the speed of an object over different time periods of its journey.

> **1** Two toy cars move round a 2 m track. Car A takes 4 seconds to complete a lap and car B takes 6 seconds. What is the speed of each car in m/s?

Figure 32.2 shows the distance/time graph for an object which moved at a steady speed (line A) then stopped and remained stationary (line B). If the object had been travelling at a higher speed, line A would be steeper. If the object had been travelling at a lower speed, line A would be less steep.

The speed of an object, that is how far the object moved in a certain time, can be calculated from the distance/time graph. For example, the object in Figure 32.2 moved at 10 cm/s.

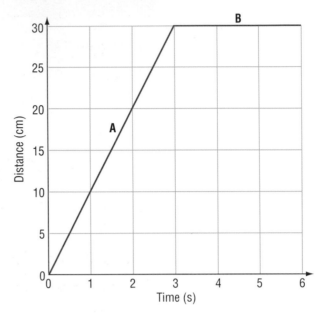

Figure 32.2   Distance/time graph.

Velocity

You will come across the word 'velocity'. This is a special word used in physics to describe both the speed and the direction of a moving object.

Speed records

Candidates trying to beat the land speed record must drive their car at full speed between two markers. Table 32.1 shows some land speed records from the end of the 20th century. You could check on the Internet to see if the 1997 record has been broken.

Figure 32.3   Breaking the land speed record in 1997.

2 How long did Art Arfan's record stand?

3 How much faster than Spirit of America was The Blue Flame?

4 By how much did the land speed record rise between 1965 and 1997?

Table 32.1 Land speed records.

Date	Speed (km/h)	Driver	Car
15/10/97	1227.99	Andy Green	Thrust SSC
4/10/83	1013.47	Richard Noble	Thrust 2
23/10/70	995.85	Gary Gabelick	The Blue Flame
15/11/65	960.96	Craig Breedlove	Spirit of America
7/11/65	922.48	Art Arfan	Green Monster

Acceleration

The acceleration of an object (moving in a straight line) is a measure of how its speed changes in a certain interval of time. The following equation shows how the steady acceleration of an object can be calculated:

$$\text{acceleration} = \frac{\text{change in speed}}{\text{time}}$$

The SI unit for acceleration is m/s/s or m/s^2. This is pronounced metres per second per second or metres per second squared.

Ways of measuring speed

The speedometer

The speedometer in a car is connected by a cable to a shaft which turns the wheels. There is a wire in the cable which is connected to the shaft by gear wheels. When the shaft turns, the wire in the cable turns too. At the other end of the wire is a magnet. It spins round when the car wheels turn. The magnet is surrounded by a circular metal cup which is affected by the magnetic field generated by the spinning magnet. The cup is made to turn, the turning effect increasing as the speed of the spinning magnet (and the moving car) increases. The cup is connected to a spring and a pointer. The spring prevents the cup spinning but allows it to turn further as the car's speed increases. The pointer turns with the cup and moves across the scale of the speedometer dial.

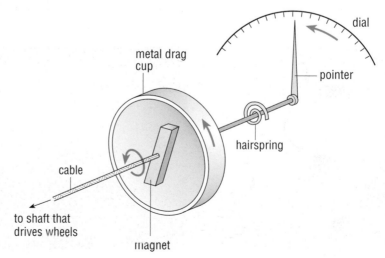

Figure 32.4 A speedometer.

The speed trap gun

The speed trap gun is a radar gun. When the gun is fired at an approaching vehicle a beam of radio waves travels to it through the air. This is reflected off the front of the vehicle and returns to a receiver on the gun. A computer in the gun compares the time difference between sending the beam and receiving it back from the vehicle and calculates the vehicle's speed.

The stop watch

For many years, the stop watch was used to measure speed. The watch was started as the speeding object passed the start line and was stopped when the object passed the finish line.

Light gates

In a light gate a beam of light shines onto a light-sensitive switch. The light gate used at the start of a speed test works in the following way. When the beam is broken by an object passing through it, the switch starts an electronic stop clock. The light gate used at the finish of the speed test causes the clock to be stopped when the beam is broken by the object passing through it.

Forces affecting speed

When objects move along a surface, friction occurs and opposes the motion. Two other forces that affect speed are air resistance and water resistance.

Air resistance

Air is a mixture of gases. When an object moves through the air it pushes the air out of the way and the air moves over the object's sides and pushes back on the object. This push on the object is called air resistance or drag.

The value of the air resistance depends on the size and shape of the object. Many cars are designed so that the air resistance is low when the car moves forwards. The car's body is designed like a wedge to cut its

5 Two people timed the speed of an object with a stop watch. They each got a slightly different result. How could this be?

6 Which is more reliable – using a manual stop watch or using light gates? Explain your answer.

Figure 32.5　Testing a streamlined sports car in a wind tunnel.

way through the air and the surfaces are curved to allow the air to flow over the sides with the minimum drag. Shapes that are designed to reduce air resistance are called streamlined shapes.

A dragster is a vehicle which accelerates very quickly. In a dragster race two vehicles accelerate along a straight track. At the end of the race the dragsters are slowed down by brakes and a parachute. The parachute offers a large surface area against which the air pushes. The high air resistance of the parachute slows down the dragster and helps it stop in a short distance.

Figure 32.6 These parachutes are used to slow down a dragster after a race.

The air resistance produced by a parachute is also used to bring sky divers safely to the ground (see pages 320–321). The resistance of the gases in the atmospheres of other planets in the Solar System is used to slow down space probes so they can land safely and the devices on board are able to carry out their investigations.

Figure 32.7 A safe landing for the Mars Exploration Rover.

7 How would the size of parachute required on a space probe to allow it to land safely differ on
a) a planet such as Venus which has a thick atmosphere,
b) a planet such as Mars which has a thin atmosphere?
Explain your answers.

Water resistance

When an object moves through water it pushes the water out of the way and the water moves over the object's sides and pushes back on the object. This push on the object is called water resistance or drag. Objects that can move through the water quickly have a streamlined shape. A fish such as a barracuda which moves quickly through the water has a much more streamlined shape than a slow-moving sunfish.

Barracuda
Figure 32.8

A sunfish

Water resistance affects the movement of ships and boats on the water surface. Boats designed for high speeds have a hull shaped to reduce water resistance as much as possible. Some boats are equipped with a device called a hydrofoil which reduces the area of contact between the boat and the water so that water resistance is kept to a minimum. The boat (itself called a hydrofoil) can then move quickly over the water surface.

A speed boat
Figure 32.9

A hydrofoil

Sky diving: terminal velocity

When a sky diver leaps from an aeroplane the diver's weight pulls him or her down. The air resistance is small compared to the weight as he or she starts to fall and so the diver accelerates (Figure 32.10a). Eventually the force of the air resistance balances the force pulling the diver towards the ground and he or she falls steadily at what is called the terminal velocity (Figure 32.10b). When the sky diver opens the parachute the air resistance greatly increases (Figure 32.10c). This slows down the sky diver to a new, slower terminal velocity so he or she can make a safe landing.

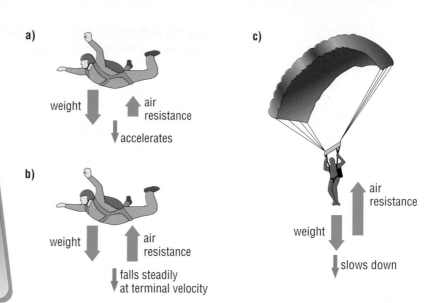

a)

weight — air resistance — accelerates

b)

weight — air resistance — falls steadily at terminal velocity

c)

air resistance — weight — slows down

Figure 32.10 The motion of a sky diver.

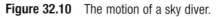

For discussion

During free fall, how does a sky diver alter the terminal velocity by altering his or her shape?

Why are tangled parachutes dangerous?

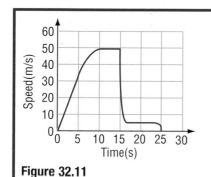

Figure 32.11

8 Figure 32.11 shows a speed/time graph of a sky diver's jump.
   a) At what time is the acceleration greatest?
   b) When did the diver begin to fall at the terminal velocity?
   c) For how long did the sky diver fall at the terminal velocity?
   d) When was the parachute opened?
   e) At what speed did the sky diver hit the ground?

Summary

- The speed of an object is the distance it travels in a certain time interval (*see page 315*).
- The acceleration of an object moving in a straight line is its change in speed in a certain time interval (*see page 317*).
- There are a variety of devices for measuring speed (*see pages 317–318*).
- Speed is affected by friction, air resistance and water resistance (*see pages 318–319*).
- When the forces on a falling object balance, it falls at its terminal velocity (*see above*).

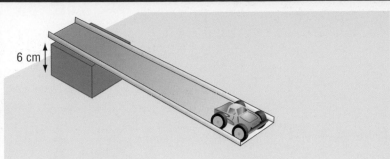

Figure 32.12

A group of pupils investigated the movement of a model car. They set up a ramp at 6 cm height and let the car roll down it and across the floor (Figure 32.12). They measured the distance travelled by the car after it left the ramp and moved across the floor. The experiment was repeated three more times with the ramp set at 6 cm, then the height was reset and more of the car's movements were recorded. Table 32.2 shows the results of the investigation.

Table 32.2

Height (cm)	Distance (cm)			
6	20	21	20	19
7	24	25	22	21
8	32	32	33	33
9	40	40	39.5	38
10	45	42	45	44
11	55	53	55	55
12	60	60	58	59
13	67	62	63	64

1 How many times was the height of the ramp changed?

2 How is an average calculated?

3 Calculate the average distance travelled for each height of the ramp.

4 Plot a graph to show the relationship between the height of the ramp and the distance travelled from the ramp by the car.

5 What conclusions can you draw from your analysis of the results of this investigation?

33 Pressure and moments

Pressure on a surface

In Chapter 26 we examined forces acting at a point on an object. In this chapter we consider the effect of a force acting over an area.

When a force is exerted over an area we describe the effect in terms of pressure. Pressure can be defined by the equation:

$$\text{pressure} = \frac{\text{force}}{\text{area}}$$

The SI unit for pressure is N/m^2 but it can also be measured in N/cm^2.

An object resting on a surface exerts pressure on the surface because of the object's weight. Weight is the force produced by gravity acting on a solid, a liquid or a gas, pulling the material downwards towards the centre of the Earth. The weight acts on the mass of that material. For example, the weight of a solid cube acts on that cube (Figure 33.1).

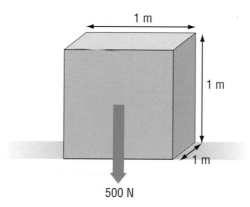

Figure 33.1 The weight acting on a cube of material.

The cube pushes down on the ground (or other surface that it rests on) with a force equal to its weight. The pressure that the cube exerts on the ground is found by using the equation above. For example, if the cube has a weight of 500 N and the area of its side is 1 m^2, the pressure it exerts on the ground is:

$$\text{pressure} = \frac{500}{1} = 500 \text{ N/m}^2$$

If the cube had a weight of 500 N and the area of its side was 2 m^2, the pressure it would exert on the ground would be:

$$\text{pressure} = \frac{500}{2} = 250 \text{ N/m}^2$$

An object exerts a pressure on the ground according to the area of its surface that is in contact with the ground. For example, a block with dimensions 1 m × 1 m × 2 m and a weight of 200 N will exert a pressure of 200/1 = 200 N/m^2 when it is stood on one end (Figure 33.2a, overleaf) but a pressure of only 200/2 = 100 N/m^2 when laid on its side (Figure 33.2b).

1 What is the pressure exerted on the ground by a cube which has a weight of 600 N and a side area of
   a) 1 m²,
   b) 3 m²?

2 What is the pressure exerted on the ground by an object which has a weight of 50 N and a surface area in contact with the ground of
   a) 1 cm²,
   b) 10 cm²,
   c) 25 cm²?

3 a) What pressure does a block of weight 600 N and dimensions 1 m × 1 m × 3 m exert when it is
      i) laid on its side,
      ii) stood on one end?
   b) Why does it exert different pressures in different positions?

4 Drivers in Iceland, when going out on the snow, let their tyres down until they are very soft. The tyres spread out over the surface of the snow as they drive along. Why do you think the drivers do this?

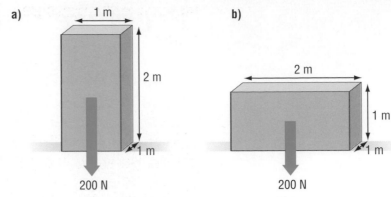

Figure 33.2   The weight acting on a block in two positions.

Your weight acting downwards causes you to exert a force on the ground through the soles of your shoes. If you lie down this force acts over all the areas of your body in contact with the ground. These areas together are larger than the areas of the soles of your shoes and you therefore push on the ground with less pressure when lying down than when you are standing up.

Figure 33.3   The force you exert downwards acts over a larger area when you lie down.

Reducing the pressure

When people wear skis the force due to their weight acts over a much larger area than the soles of a pair of shoes. This reduces the pressure on the soft surface of the snow and allows the skier to slide over it without sinking.

Figure 33.4   Skis stop you sinking into the snow.

Increasing the pressure

Studs

Sports boots for soccer and hockey have studs on their soles. They reduce the area in contact between your feet and the ground. When you wear a pair of these boots your downward force acts over a smaller area than the soles of your feet and you press on the ground with increased pressure. Your feet sink into the turf on the pitch and grip the surface more firmly. This makes it easier to run about without slipping while you play the game.

Figure 33.5 The studs on this soccer boot help the player to grip the turf.

Pins and spikes

When you push a drawing pin into a board the force of your thumb is spread out over the head of the pin so the low pressure does not hurt you. The same force, however, acts at the tiny area of the pin point. The high pressure at the pin point forces the pin into the board.

Sprinters use sports shoes which have spikes in their soles. The spike tips have a very small area in contact with the ground. The weight of the sprinter produces a downward force through this small area and the high pressure pushes the spikes into the hard track, so the sprinter's feet do not slip when running fast.

5 A girl wearing trainers does not sink into the lawn as she walks across it but later when she is wearing high-heeled shoes she sinks into the turf. Why does this happen?

Figure 33.6 The spikes stop the sprinter from slipping on the track.

Knives

As we have seen, high pressure is made by having a large force act over a small area. The edge of a sharp knife blade has a very small area but the edge of a blunt knife blade is larger. If the same force is applied to each knife the sharp blade will exert greater pressure on the material it is cutting than the blunt knife blade and will cut more easily than the blunt blade.

Figure 33.7 Knives cut well when they are sharp because of the high pressure under the blade.

Pressure in a liquid

Matter is made from particles. In solids the particles are held in position. In liquids the particles are free to slide over each other, and in gases the particles are free to move away from each other. A full description of particles in matter is given in Chapter 14.

In a solid object the pressure of the particles acts through the area in contact with the ground. In a liquid the pressure of the particles acts not only on the bottom of the container but on the sides too (Figure 33.8).

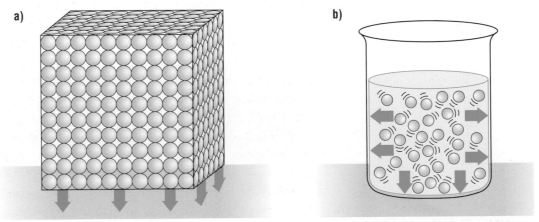

Figure 33.8 Pressure exerted by **a)** particles in a solid block and **b)** particles in a liquid.

Pressure and depth of a liquid

The change in pressure with depth in a liquid can be demonstrated by setting up a can as shown in Figure 33.9. When the clips are removed from the three rubber tubes, water flows out as shown.

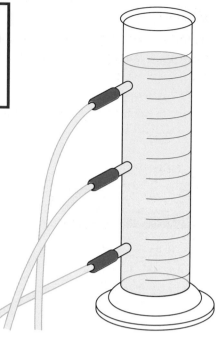

6 How does the path of the jet of water at the bottom of the can in Figure 33.9 change as the water level in the can falls? Why does it change?

All three jets of water leave the can horizontally but the force of gravity pulls them down. The water under the greatest pressure travels the furthest horizontally before it is pulled down. The water under the least pressure travels the shortest distance horizontally before it is pulled down.

Figure 33.9 Jets of water leaving a can at different depths.

A closer look at pressure and depth

The mass of a cubic centimetre of water is 1 g or 0.001 kg. The force of gravity (10 N/kg) means that this mass exerts a force downwards. The size of the force (equal to its weight) is calculated by:

$$0.001 \times 10 = 0.01 \text{ N}$$

This force acts on an area of 1 cm^2 (Figure 33.10a) so the pressure it exerts is:

$$\frac{0.01}{1} = 0.01 \text{ N/cm}^2$$

If a second cube of water is placed over the first, the pressure beneath the lower cube is increased to 0.02 N/cm^2 since the weight of water has doubled but the area it rests on has not (Figure 33.10b).

In fact, the pressure exerted by a liquid depends on the height of the column of liquid above its base, no matter what the area of the base of the column. Consider four cubes of water placed as in Figure 33.10c. A force of 0.04 N acts over an area of 4 cm^2 so the pressure is 0.01 N/cm^2, as in Figure 33.10a.

If water is placed in the two arms of a vessel as shown in Figure 33.11 (overleaf) and the partition between the arms is removed, the water moves down the left arm and up the right arm until the water in both arms is at the same level. When this happens both columns of water are exerting the same pressure on the bottom of the vessel.

a)

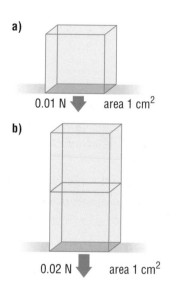

0.01 N area 1 cm^2

b)

0.02 N area 1 cm^2

c)

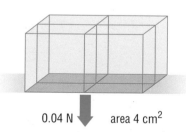

0.04 N area 4 cm^2

Figure 33.10 The pressure doubles when the depth of water doubles, but the pressure does not depend on the area of the column of water.

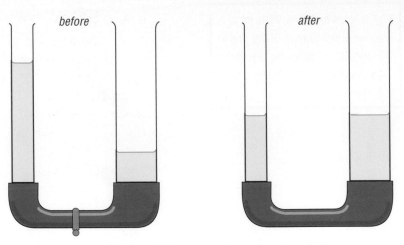

Figure 33.11 Water flows until the pressure in each column is the same.

You can see that although the arms are of different widths the water in them settles to the same level in each. The wider arm has more of the liquid in it but it also has a larger area. The water level indicator on some jug kettles uses this fact to allow you to see where the water level is inside (Figure 33.12).

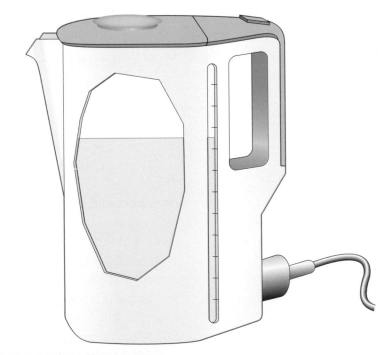

Figure 33.12 Inside a jug kettle.

Hydraulic equipment

If pressure is applied to the surface of a liquid in a container, the liquid is not squashed. It transmits the pressure so that pressure pushes on all parts of the container with equal strength.

In hydraulic equipment a liquid is used to transmit pressure from one place to another. The pressure is applied in one place and released in another. If the area where the pressure is applied is smaller than the area where the pressure is released, the strength of the force is increased as the following example shows.

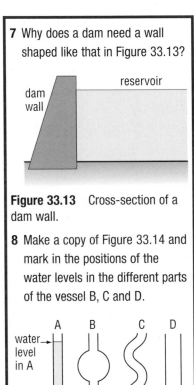

7 Why does a dam need a wall shaped like that in Figure 33.13?

Figure 33.13 Cross-section of a dam wall.

8 Make a copy of Figure 33.14 and mark in the positions of the water levels in the different parts of the vessel B, C and D.

Figure 33.14 Pascal's vases, named after Blaise Pascal (1623–1662).

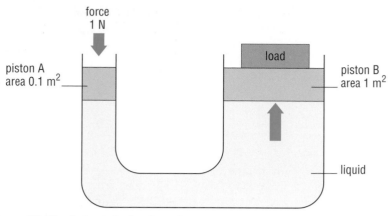

Figure 33.15 A simple hydraulic system.

A force of 1 N is exerted on area A of 0.1 m² (Figure 33.15). The pressure exerted on this is:

$$\frac{1}{0.1} = 10 \text{ N/m}^2$$

This same pressure is transmitted throughout the liquid and acts over area B. Area B is 1 m².

The equation pressure = force/area can be rearranged to find the force:

force = pressure × area

Using this rearranged equation the force at B can be found:

force = 10 N/m² × 1 m² = 10 N

The force has been increased from 1 N to 10 N.

A car may be raised with a small force by using a hydraulic jack. When a small force is applied to a small area of the liquid in the jack, a larger force is released across a larger area and acts to raise the car.

Figure 33.16 This car has been raised into the air for repairs by a hydraulic jack.

9 Why are hydraulic systems known as 'force multipliers'?

The brake system on a car is a hydraulic mechanism. The small force exerted by the driver's foot on the brake pedal is converted into a large force acting at the brake pads. This results in a large frictional force that makes it harder for the wheels to turn and so stops the car.

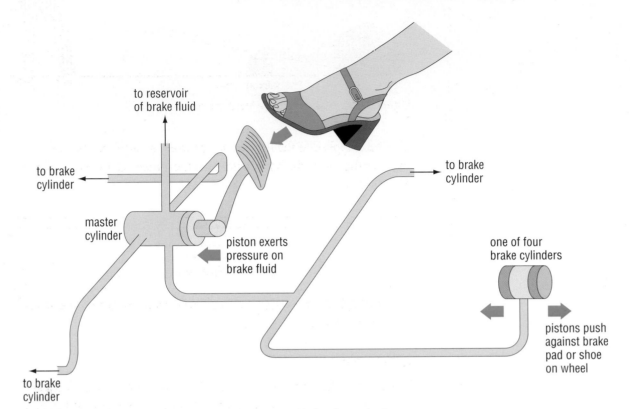

Figure 33.17 Hydraulic car brakes.

The turning effect of forces

A force can be used to turn an object in a circular path. For example, when you push down on a bicycle pedal the cog wheel attached to the crankshaft turns round.

A nut holding the hub of a bicycle wheel to the frame is turned by attaching a spanner to it and exerting a force on the other end of the spanner in the direction shown in Figure 33.18.

Levers

A device that changes the direction in which a force acts is called a lever. It is composed of two arms and a fulcrum or pivot. The lever also acts as a force multiplier. This means that a small force applied to one arm of the lever can cause a large force to be exerted by the other arm of the lever. For example, a crowbar is a simple lever. It is used to raise heavy objects. One end of the crowbar is put under a heavy object and the crowbar is rested on the fulcrum, as in Figure 33.19. When a downward force is applied to the long arm of the crowbar an upward force is exerted on the heavy object. A small force acting downwards at a large distance from the fulcrum on one side produces a large force acting upwards a short distance from the fulcrum on the other arm.

Figure 33.18 Tightening a nut.

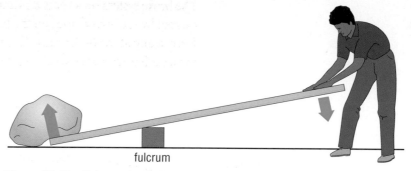

Figure 33.19 Using a simple lever: a crowbar.

The force applied to the lever to do the work is called the effort. It opposes the force which is resisting the movement, called the load. See Figure 33.20.

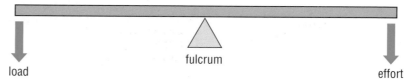

load   fulcrum   effort

Figure 33.20 A simple lever: the load at one end is overcome by the effort at the other end.

Moments

The turning effect produced by a force around a fulcrum is called the moment of the force. The direction of the moment is usually specified as clockwise or anticlockwise about the fulcrum. The size of the moment is found by multiplying the size of the force by the distance between the point at which the force acts and the fulcrum. The moment of a force can be shown as an equation:

$$\text{moment of force} = \text{force} \times \text{distance from the fulcrum}$$

The moment is measured in newton-metres (N m). The moment of the force applied to one arm of a lever is equal to the moment of the force exerted by the other arm. For example, a 100 N force applied downwards 2 m from the fulcrum on one arm produces a 200 N force upwards 1 m from the fulcrum on the other arm.

In the case of a see-saw, which is another simple type of lever, the moment of the weight on one arm must equal the moment of the weight on the other arm for the see-saw to balance.

A pair of pliers (see Figure 33.21) is made from two levers. When they are used to grip something a small force applied to the long handles produces a large force at the short jaws.

10 What is the moment of a 100 N force acting on a crowbar
   a) 2 m from the fulcrum,
   b) 3 m from the fulcrum,
   c) 0.5 m from the fulcrum?

11 A 100 N force acting on a lever 2 m from the fulcrum balances an object mass 0.5 m from the fulcrum on the other arm. What is the weight of the object (in newtons)? What is its mass (in kg)?

12 Where will the strongest force be exerted by scissor blades to cut through a piece of material? Explain your answer.

13 Why can a lever be described as a force multiplier?

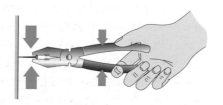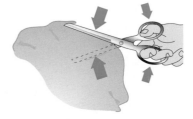

Figure 33.21 Using pliers and scissors.

Summary

- Pressure acts when a force acts over an area of surface (*see page 323*).
- When a solid object exerts a pressure on the surface below it, the smaller the area of contact, the greater the pressure (*see page 325*).
- Pressure in a liquid acts in all directions and increases with the depth of the liquid (*see pages 326–328*).
- In hydraulic systems pressure is transmitted through a liquid (*see pages 328–330*).
- A force can produce a turning effect (*see page 330*).
- A lever is a device that changes the direction in which a force acts (*see page 330*).
- The turning effect produced by a force around a fulcrum is called a moment (*see page 331*).

End of chapter question

1 Is it possible to balance a mass of weight 5 N with a mass of weight 15 N on a model see-saw with 10 cm arms? Explain your answer. Suggest where you might place each mass to get an exact balance.

Glossary

A

acceleration For an object moving in a straight line, the change in speed in a certain period of time.

acid A substance with a pH less than 7 that reacts with metals to produce hydrogen.

acid rain Rain produced by the reaction of sulphur dioxide and oxides of nitrogen with water in clouds. It has a pH of less than 5.

adaptation The way a living thing is suited to its habitat so that it can survive there. Adaptation can also mean the process by which living things become more suited to their habitat.

addiction A condition in which a person is unable to lead a normal life without taking drugs or alcohol on a regular basis.

adolescence The time in a person's life when they change from a child to an adult.

aerobic respiration The release of energy from food using oxygen.

air resistance The backward push of the air on an object moving through it.

alchemy The ancient study of chemical reactions to produce gold from less expensive metals, or to produce a chemical that would extend life.

alimentary canal The digestive tube that begins with the mouth and ends with the anus. It is also sometimes called the gut.

alkali A base that is soluble in water and makes an alkaline solution.

alkaline The condition of a solution in which the pH is greater than 7.

allotrope One of two or more forms in which an element can exist. For example, carbon can exist as diamond or graphite.

alloy A mixture of two or more metals, or of a metal such as iron with a non-metal such as carbon.

amino acid A molecule containing carbon, hydrogen, oxygen and nitrogen. It links up with other amino acids to form long-chain molecules called proteins.

ammeter An instrument for measuring the size of a current flowing through a circuit, in amperes.

amnion A sac that surrounds the embryo that is filled with a watery fluid.

amplitude The maximum displacement of a vibrating object from its rest position.

anaemia An unhealthy condition that may be due to the lack of iron in the diet. One of the symptoms is tiredness.

anaerobic respiration The release of energy from food without the use of oxygen.

antagonistic muscles A pair of muscles in which each of the contracting muscles brings about a movement that is opposite in direction to the other.

anther The organ in a flower that produces pollen grains.

antibiotic A chemical made by some microbes or produced artificially by chemical reactions that is used to kill certain kinds of disease-causing bacteria.

antibody A chemical made by some white blood cells to protect the body from disease-causing microbes and their toxins.

antigen A feature found on the body of a disease-carrying microbe that stimulates the human body to produce antibodies.

artery A blood vessel with elastic walls that carries blood away from the heart.

asexual reproduction The process of producing offspring without the making of gametes and the process of fertilisation.

atom The smallest particle of an element that can take part in a chemical reaction.

B

base A substance that can take part in a chemical reaction with an acid, forming a salt and water.

battery Two or more electrical cells joined in series in a circuit.

bile A substance made by the liver and stored in the gall bladder. It is released onto food in the duodenum to aid the digestion of fats.

biodegradable The property of a complex substance that allows it to be broken down into simple substances by the action of decomposers.

biogas Methane produced by the digestive processes of microorganisms that feed on plant and animal waste.

biomass The mass of an organism or group of organisms after their bodies have been dried out.

biotechnology The use of biological processes to make useful substances, such as antibiotics, and to produce new kinds of living organisms through genetic engineering.

boiling A process in which a liquid turns to a vapour at the liquid's boiling point.

boiling point The highest temperature to which a liquid can be heated before the liquid turns into a gas.

C

capillary A blood vessel with one-cell thick walls through which substances pass between the blood and the surrounding cells.

carbohydrate A nutrient made from carbon, hydrogen and oxygen. Most are made by plants.

carnivore An animal that only eats other animals for food.

carpel The female organ of a flower that produces the fruit and the seed.

catalyst A substance that speeds up a chemical reaction without being used up in the reaction.

cell (in biology)   The basic unit of life. The cell contains a nucleus, cytoplasm and membrane around the outside. The bodies of most living things are made from large numbers of cells.

cell (in physics and chemistry)   A device containing chemicals which react and produce a current of electricity in a closed conducting circuit.

chemical potential energy   The energy stored in the links between atoms of a substance.

chlorophyll   A green pigment found mainly in plant cells that traps energy from sunlight and makes it available for the process of photosynthesis.

chloroplast   A component of a cell. It is green and absorbs some of the energy of sunlight for use in photosynthesis.

chromatography   A process in which substances dissolved in a liquid are separated from each other by allowing the liquid to flow through porous paper.

chromosome   A thread-like structure that appears when the cell nucleus divides. It contains DNA.

cilia   Short hair-like projections that may form on the surface of a cell. They can beat to and fro to move the bodies of Protoctista or to help with the movement of fluids in animal systems.

circuit breaker   An electrical device that stops the flow of electricity through a circuit if the current becomes too high to pass safely.

clone   One of a number of identical individuals produced by asexual reproduction.

combustion   A chemical reaction in which a substance combines with oxygen quickly and heat is given out in the process. If a flame is produced, burning is said to take place.

compound   A substance made from the atoms of two or more elements that have combined by taking part in a chemical reaction.

condensation   A process in which a gas cools and changes into a liquid.

conduction (electrical)   The passage of electricity (moving charge) through a material.

conduction (thermal)   The passage of heat energy from one part of a material to another, by vibrating particles passing kinetic energy on to neighbouring particles.

conductor (electrical)   A material that allows electricity to pass through it.

conductor (thermal)   A material that allows heat to pass easily through it by conduction.

constellation   A pattern of stars in the sky which appear to form a group but may in fact be light years apart.

consumer   An animal that eats either plants or other animals.

convection   The passage of heat energy through a liquid or a gas, by the particles in the substance changing position.

cross-pollination   The transfer of pollen from the anthers of a flower on one plant to the stigma of a flower on another plant of the same species.

crystal   A substance made from an orderly arrangement of atoms or molecules that produces flat surfaces, arranged at certain angles to each other.

crystallisation   A process in which crystals are formed from a liquid or a gas.

cytoplasm   A fluid-like substance in the cell in which processes take place to keep the cell alive.

D

decant   A process of separating a liquid from its sediment by pouring the liquid away from the sediment.

decomposition   A chemical reaction breaking down a substance into simpler substances.

density   The mass of a substance that is found in a certain volume.

diffusion   A process in which the particles in two gases or two liquids, or the particles of a solute in a solvent, mix on their own without being stirred.

digestion   The process of breaking down large food particles into small ones so that they can be absorbed by the body.

diode   An electrical component which allows a current to pass in only one direction through it.

dispersion   The spreading out of light of different colours from a beam of sunlight.

displacement   A reaction in which a metal in a salt is replaced by another metal.

distillate   A liquid produced by distillation.

distillation   A process of separating a solute from a solvent by heating the solution they make, until the solvent turns into a gas and is condensed and collected separately without the solute.

DNA (deoxyribonucleic acid)   A substance in the nuclei of cells that contains information, in the form of a code, about how an organism should develop and function.

drag   A force which acts on a moving body in the opposite direction to which the body is moving, due to air or water resistance.

dynamo   A device for generating a current of electricity. There are two kinds: a direct current or d.c. dynamo which produces a current that flows in only one direction, and an alternating current or a.c. dynamo which produces a current that changes direction many times a second. The bicycle dynamo and power station generators are a.c. dynamos.

E

ecology   The study of living things in their natural surroundings or habitat.

ecosystem   An ecological system in which the different species in a community react with each other and with the non-living environment. Ecosystems are found in all habitats such as lakes and woods.

egestion (*see also* excretion)   The release of undigested food and other contents of the alimentary canal from the anus.

elastic limit   The maximum force that can be applied to an elastic material without the material becoming permanently deformed.

elastic material   A material that exerts a strain force when deformed, tending to restore it to its original shape.

electrolysis   The process in which a chemical decomposition occurs due to the passage of electricity through an electrolyte.

electrolyte   A solution or molten solid through which a current of electricity can pass.

electromagnet   A magnet which is made by coiling a wire around a piece of iron then passing a current through the wire.

electromagnetic waves   Waves with electrical and magnetic properties which transfer energy, such as light, infrared and radio waves.

electron   A tiny particle in an atom which moves round the nucleus. It has a negative electric charge.

electrostatic charge   A charge of electricity which stays in place on the surface of a material; it may be positive (due to a lack of electrons) or negative (due to an excess of electrons).

element   A substance made of one type of atom. It cannot be split up by chemical reactions into simpler substances.

embryo   The body of an organism in its early development from a zygote (*see also* zygote). In humans an embryo develops in the womb in the first two months of pregnancy (*see also* fetus).

energy   A measure of the work that has been done or is able to be done.

energy chain   The flow of energy through one or more energy transducers.

energy transducer   A material or an object in which energy changes from one form to another as it passes through; also called an energy converter.

endoskeleton   A skeleton on the inside of the body, as occurs in vertebrates.

enzyme   A chemical made by a cell that is used to speed up chemical reactions in life processes such as digestion and respiration.

evaporation   A process in which a liquid turns into a gas without boiling.

evolution   The process by which one species of organism is believed to change genetically over a period of time.

excretion (*see also* egestion)   The release of waste products made by chemical reactions inside the body.

exoskeleton   A skeleton on the outside of the body, as occurs in arthropods.

F

fats   Food substances that provide energy. They belong to a group of substances called lipids, which includes oils and waxes.

fermentation   A type of anaerobic respiration that occurs in yeast and bacteria. Some fermentation processes are used to make alcohol.

fertilisation   The fusion of the nuclei from the male and female gametes that results in the formation of a zygote.

fetus   A stage in the development of the mammal in the womb when the main features of the animal have formed. In humans the fetus develops in the womb from the second to the ninth month of pregnancy.

field   A region in which a non-contact force acts.

filtration   A process of separating solid particles from a liquid by passing the liquid through paper with small holes in it.

force   A push or a pull; it may be a contact force, for example an impact force, or a non-contact force, for example a magnetic force.

fossil fuel   A fuel such as coal, oil or gas which is formed from the fossilised remains of plants or animals.

frequency   The number of waves passing a point in a certain amount of time.

friction   A force that acts against the relative movement of two surfaces in contact.

fruit   A structure that forms from the ovary of a flowering plant after fertilisation has taken place.

fuel   A substance used to provide energy for heating, producing electricity or working machinery.

fulcrum   The point (pivot) on which a lever is supported as it turns.

fuse   A wire with a low melting point which is placed in a circuit to prevent high currents flowing through it.

G

galaxy   A large group of stars held together by gravitational forces.

gamete   A cell involved in sexual reproduction, i.e. a sperm or egg cell in animals.

gas   A substance whose volume changes to fill any container into which it is poured.

gene   A section of DNA that contains the information about how a particular characteristic, such as hair colour or eye colour, can develop in the organism.

genetic engineering   The process of moving genes between different types of organisms to produce new organisms with particularly useful properties.

geothermal energy   Energy extracted from hot rocks beneath the surface of the Earth.

germination   The process in which the plant inside a seed begins to grow and bursts out of the seed coat.

global warming   The raising of the temperature of the atmosphere due to the greenhouse effect.

gravitational force   A force of attraction between any two masses in the Universe. The force is noticed when the two masses are very large, for example planets, or when one is very large and the other is very small by comparison, for example the Earth and you.

gravitational potential energy   The energy stored in an object because of its position above the Earth's surface.

greenhouse effect   The trapping in the atmosphere of the Sun's heat that is reflected from the Earth's surface.

growth hormone   A chemical produced by the pituitary gland in the head. It makes the body grow.

H

habitat   The place where a particular living thing survives.

haemoglobin   The pigment in red blood cells that contains iron and transports oxygen around the body.

heat (thermal) energy   The energy transferred to or from a substance by heating, which increases or decreases the internal kinetic energy of the substance.

herbivore   An animal that eats only plants for food.

hormone   A chemical, secreted by a gland in the body, which travels in the blood and acts on particular parts of the body. It may produce changes in growth or activity.

hydraulic system   A machine made from pistons and pipes that contain a liquid. It transmits pressure and converts a small force into a large one.

hydrocarbon   A compound made from hydrogen and carbon only.

hydroelectric power   Electricity produced from the energy of falling water.

hygiene   The study and practice of maintaining health by keeping the body clean.

I

image   The picture of an object which is produced when light is reflected from a mirror or is focused onto a screen by a lens.

immiscible   A property of a liquid that does not allow it to mix with another liquid.

immunisation   A process in which the body is made resistant or immune to a disease.

impact force   The force exerted by one object on another when they collide.

incandescence   The glowing of a substance, due to the amount of heat that it has received.

incubation   A process in which organisms, such as a developing chick in an egg or colonies of bacteria, are kept at a constant, raised temperature to aid their growth and development.

inertia   The opposition of an object to a change in its motion.

infrared radiation   Waves of electromagnetic radiation with wavelength between red light and microwaves; they cause warming.

insulator (electrical)   A material that does not allow electricity to pass through it.

insulator (thermal)   A material that does not allow heat to pass easily through it by conduction.

internal energy   The energy that atoms of a substance possess, partly due to their motion.

invertebrate   An animal that does not have a skeleton of cartilage or bone inside its body.

J

joint   A place where two bones meet. In movable joints the bones are held together by ligaments and are capped in cartilage to reduce friction.

K

kinetic energy   The energy of a moving object.

L

LDR   Light dependent resistor.

LED   Light emitting diode.

lens   A piece of transparent glass or plastic with at least one curved surface, used to change the direction of light rays by refraction.

lever   A device that changes the direction in which a force acts, for example a crowbar.

light year   The distance travelled by light through space in one year. It is 9.5 million million kilometres.

liquid   A substance with a definite volume that flows and takes up the shape of any container into which it is poured.

luminous object   Any object that releases energy in the form of light.

lymphocyte   A white blood cell that makes antibodies to destroy bacteria.

M

magnetic material   A material that is attracted to a magnet and can be made into a magnet.

magnetic pole   One of two regions in a magnet where the magnetic force is very strong.

mass   The amount of matter in a substance or object. It is measured in units such as grams and kilograms.

menopause   The time in a woman's life, usually about the age of 50, when monthly periods (menstruation) stop.

menstruation   A period of time each month when the uterus loses the lining of its wall.

metal   A member of a group of elements which are shiny, good conductors of heat and electricity, and malleable (bendy).

milk teeth   The first set of teeth that grow in the jaws of young mammals, including humans.

mineral (in biology)   A substance taken up from the soil water by the plant roots and used for growth and development of the plant. It is also an essential nutrient in the diet of animals.

mineral (in chemistry)   A substance that has formed from an element or compound in the Earth and exists separately, or with other minerals to form rocks.

miscible   A property of a liquid that allows it to mix freely with another liquid.

molecule   A group of atoms joined together that may be identical, in the molecules of an element, or different, in the molecules of a compound.

moment   The turning effect of a force.

N

natural selection   The process by which evolution is thought to take place. Individuals in a species best suited to an environment will thrive there and produce more offspring, while less well suited individuals will produce fewer offspring. In time the less well suited will die out leaving the best suited individuals to form a new species.

neutralisation   A reaction between an acid and a base in which the products (a salt and water) do not have the properties of the reactants.

neutron   A particle in the nucleus of an atom which has no electrical charge.

non-luminous object   An object that does not release energy in the form of light but may reflect light from luminous objects.

non-magnetic material   A material that is not attracted to a magnet and cannot be made into a magnet.

non-metal A member of a group of elements that are not shiny or malleable, and most do not conduct heat or electricity.

non-renewable energy source A source of energy, such as fossil fuels and radioactive materials, which cannot be replaced once it has been used.

nuclear energy The energy stored in the nucleus of an atom.

nuclear fission The process in which the nucleus of an atom breaks down into smaller nuclei and releases energy.

nuclear fusion A process in which atomic nuclei join together to form larger nuclei of other elements.

nuclear reactor A device in which nuclear fission is allowed to take place safely so that the energy released can be used to generate electricity.

nucleus (in biology) The part of the cell that contains the DNA and controls the activities and development of the cell.

nucleus (in physics and chemistry) The central part of an atom, which contains particles called protons and neutrons.

nutrient A substance in a food that provides a living thing with material for growth, development and good health.

O

omnivore An animal that eats both plants and animals for food.

opaque material A material through which light cannot pass.

ore A rocky material that is rich in a mineral from which a metal can be extracted.

organ A part of an organism, made from a group of cell tissues, that performs an important function in the life of the organism.

ovary The organ where the female gametes are made in plants and animals.

oxidation The chemical combination of a substance with oxygen to form an oxide.

P

periodic table The arrangement of the elements in order of their atomic number that allows elements with similar properties to be grouped together.

peristalsis The wave of muscular contraction that moves food along the alimentary canal.

phenomenon Something which can be observed that is due to the way matter behaves; for example when a piece of wood gets very hot it bursts into flame and gives out light.

phloem A living tissue in a plant through which food made in the leaves passes to all parts of the plant.

photosynthesis The process by which plants make carbohydrates and oxygen from water and carbon dioxide, using the energy from light that has been trapped in chlorophyll.

pitch A measure of the frequency of a sound wave.

placenta A disc of tissue that is connected to the uterus wall and supplies the baby with oxygen and food from the mother's blood and releases waste from the baby into the mother's circulatory system.

plankton Very small (including microscopic) organisms that live near the water surface in large aquatic environments such as oceans and lakes.

pollen Microscopic grains produced by the anther, which contain the male gamete for sexual reproduction in flowering plants.

pollination The transfer of pollen from an anther to a stigma.

potential difference The difference in electric potential between two points of a circuit, which is the cause of the current flow. The current flows from a point of higher potential to a point of lower potential.

potential energy Energy that is stored when work is done moving an object against a force. It is the energy something has due to its position, for example gravitational potential energy: a person who has climbed to the top of a slide has a larger amount of potential energy than someone who has just slid to the bottom.

power A measure of the rate at which a device does work.

precipitate Particles of a solid that form in a liquid or a gas as a result of a chemical reaction.

precipitation The process of forming a precipitate in a liquid or a gas.

pressure A measurement of a force that is acting over a certain area.

prism A piece of transparent glass or plastic with a triangular cross-section, used to disperse the coloured light in sunlight to form the visible spectrum.

products The substances that are produced when a chemical reaction takes place.

protein A substance made from amino acids. Proteins are used to build many structures in the bodies of living things.

proton A particle in the nucleus of an atom that has a positive electrical charge.

puberty The time of body growth in humans during which the reproductive organs become fully developed.

R

radiation A form of energy transfer by electromagnetic waves.

radioactive materials Materials in which nuclear reactions take place and energy is released in the form of nuclear radiation.

reactants The substances that take part in a chemical reaction.

reaction force If object A exerts a force on object B, object B exerts an equal and opposite reaction force on object A.

reactivity series The arrangement of metals in order of their reactivity with oxygen, water and acids, starting with the most reactive metal.

real image An image that can be focused onto a screen.

reduction The removal of oxygen from a compound by another substance.

reflection A process in which light rays striking a surface are turned away from the surface.

refraction The bending of a light ray as it passes from one transparent substance to another.

relay A type of switch used in a low-current circuit to control the current in a second, high-current circuit.

renewable energy source A source of energy such as sunlight, wind and biomass, which can be used again and again.

resistance (electrical) The property of a material which opposes the flow of a current through the material.

resistor A device that offers a certain amount of resistance to a current passing through a circuit.

respiration The process in which energy is released from food.

reversible reaction A chemical reaction which can be reversed. The products of the forward reaction become the reactants of the reverse reaction.

S

saliva A watery substance produced by glands in the mouth that makes food easier to swallow and begins the digestion of carbohydrates.

salt A compound that is formed when an acid reacts with a substance such as a metal, or when an acid reacts with a base.

satellite A device that moves in orbit around the Earth (or any moon around any planet).

sediment A collection of solid particles that settle out from a mixture of a solid and a liquid.

seed A structure that forms from the ovule after fertilisation. It contains the embryo plant and a food store.

self-pollination The transfer of pollen from the anthers to the stigma of the same flower.

sepal A leaf-like structure that protects a flower when it is in bud.

sexual reproduction The form of reproduction in which gametes are formed, fertilisation takes place and a zygote is formed.

sliding friction The friction that exists between two objects when one is moving over the other.

solar cell A device that converts the energy in sunlight into electrical energy.

solar panel A device for collecting heat from the Sun to warm water. (Or an array of solar cells.)

Solar System The Sun and the planets, moons, asteroids and comets which move around it.

solid A substance that has a definite shape and volume.

soluble A property of a substance that allows it to dissolve in a solvent.

solute A substance that can dissolve in a solvent.

solution A liquid that is made from a solute and a solvent.

solvent A liquid in which a solute can dissolve.

sound energy The energy transferred by a sound wave.

space probe An unmanned spacecraft for exploring space and the objects in the Solar System.

spectrum (electromagnetic) The full range of wavelengths of electromagnetic waves from the shortest gamma rays to the longest radio waves.

spectrum (visible) The bands of coloured light seen when a prism disperses sunlight. The colours are red, orange, yellow, green, blue, indigo and violet.

speed A measure of the distance covered by a moving object in a certain time.

spore A reproductive structure, containing one or more reproductive cells, produced by fungi and plants that do not produce seeds, such as mosses and ferns. They are also produced by bacteria so that they can survive harsh environmental conditions.

stamen A structure in a flower composed of the anther and the filament.

standard form The standard system of recording large numbers, with only one figure in front of the decimal point, for example 3.84×10^8.

static friction The friction that exists between two objects when there is no movement between them. It acts against an applied force, preventing movement.

stigma The region on a carpel on which pollen grains are trapped.

strain force The force exerted by an elastic material when it is deformed; it acts in the opposite direction to the applied force.

strain potential energy The energy stored in an elastic material when it is deformed.

streamlined shape A shape that allows an object to move easily through air or water.

sublimation A process in which a solid turns into a gas, or a gas turns into a solid. There is no liquid stage in this process.

suspension A collection of tiny, solid particles that are spread out through a liquid or a gas.

synthesis A chemical reaction in which a substance is made from other substances.

T

temperature A measure of the hotness or coldness of a substance; it depends on the average kinetic energy of the particles.

terminal velocity The speed at which an object falls through air when the air resistance balances the weight of the object.

testes The male reproductive organs in animals. They produce sperm.

tissue A structure made from large numbers of one type of cell.

toxins Poisons produced by bacteria which cause disease.

translucent material A material that allows some light to pass through it but scatters this light in all directions.

transparent material A material that allows light to pass through it without the light being scattered.

U

ultraviolet Electromagnetic waves with wavelength shorter than blue light.

unit A standard for measurements, for example the kilogram.

upthrust The upward force exerted on an object by the liquid or gas around it that it displaces. The force is equal to the weight of the displaced liquid or gas.

ureter The tube connecting the kidney to the bladder, through which urine flows.

urethra The tube connecting the bladder to the outside, through which urine passes. In males sperm also pass along this tube.

urine　A watery solution that contains urea, the main nitrogenous waste product of humans.

uterus　The female organ in which the embryo and fetus develop. It is also known as the womb in humans.

V

vaccine　A substance that promotes the production of antibodies to protect the body from certain diseases.

vacuole　A large cavity in a plant cell that is filled with a watery solution called cell sap. May also occur as small, fluid-filled cavities in some animal cells and some Protoctista.

vacuum　A space in which there is no matter; it contains no particles.

variation　A feature that varies among individuals of the same species, such as height or hair colour.

vein　A thin-walled blood vessel that transports blood towards the heart.

vertebrate　An animal that has a skeleton inside its body made of cartilage or bone.

vibration　The rapid movement of an object to and fro about a rest position, as seen when a guitar string has been plucked.

virtual image　An image such as the one seen in a mirror which cannot be focused onto a screen.

vitamin　A substance made by plants or animals that is an essential component of the diet to keep the body in good health.

voltage　The difference in electric potential between two points such as the terminals of a cell, measured in volts.

W

weathering　The breaking up of a rock due to the action of water or air.

weight　The gravitational force between an object on a planet such as the Earth and the Earth itself. The weight pulls the object towards the centre of the Earth.

work　The energy expended when a force moves an object through a distance.

X

X-rays　Electromagnetic waves with wavelength between ultraviolet light and gamma rays.

xylem　The non-living tissue in a plant through which the water and minerals pass from the root through to the shoot.

Z

zygote　The cell produced after fertilisation has occurred. It divides, grows and eventually forms a new individual.

Index